THE YEAR IN TREES

The Year in Trees

Superb Woody Plants
for
Four-Season Gardens

KIM E. TRIPP AND J. C. RAULSTON

TIMBER PRESS
Portland, Oregon

Photograph Credits
All photographs are by J. C. Raulston except where noted in captions. The authors gratefully
acknowledge contributing photographers Robert Hays, Robert Hyland, Claire Sawyers, and
Wayside Gardens Company.

ISBN 0-88192-320-6

Printed in China

TIMBER PRESS, INC.
The Haseltine Building
133 S.W. Second Avenue, Suite 450
Portland, Oregon 97204, U.S.A.

Tripp, Kim E.
     The year in trees : superb woody plants for four-season gardens / Kim E. Tripp and
J.C. Raulston.
          p.   cm.
     Includes index.
     ISBN 0-88192-320-6
     1. Ornamental trees, 2. Ornamental shrubs. 3. Ornamental climbing plants. 4. Seasons.
I. Raulston, J. C. II. Title.
SB435.T76   1995
635.9'96—dc20
                                                                    95-3999
                                                                    CIP

Contents

Color plates follow pages 32, 96, 144, and 176

Preface

by Kim E. Tripp

This is a book for people who love plants. In particular, it is a book for people who love woody plants—trees, shrubs and vines. In the pages of this book you will find a seasonal series of woody plant portraits, a garden gallery of old friends and new acquaintances, including many unusual plants. Each portrait will extol the virtues of a plant, give cultural and propagation information, and offer advice on creative ways to use the plant in your garden. The array of plants described includes 150 wonderful woodies: a range of plants that offer unique garden character, and among which are plants useful for gardens in diverse areas.

This collection of plant portraits has been gathered from a weekly series of plant portraits I wrote as individual press releases while I worked at the North Carolina State University Arboretum in Raleigh, North Carolina, from 1990–1993. The weekly press releases, featuring a different plant each week, were designed to focus attention on plants of promise at the NCSU Arboretum currently new to, or underutilized by, the gardening public. The press release program was the brainchild of Catherine Knes-Maxwell, the Arboretum's development director—a good friend to both plants and plant-people.

Each of these plant portraits was written as I fell in love with a given plant, usually during the season of year in which the plant was at its best—but don't be surprised to find an evergreen juniper described in summer and deciduous alders chosen for winter. Woody plants are four-season plants with characteristics of interest year-round—and there was no predicting which plant I might fall in love with at any given time of the year.

The choice of which plant to describe each week was purely a matter of my personal obsession that week. As a result, the plants covered in this book are a unique collection of woody plants that range from commonly available but often overlooked selections, to rare exotics coveted by collectors, to fantastic finds brand new to horticulture. There is, however, one common thread throughout: all of these plants are "good garden plants"; i.e., they are especially appealing, relatively trouble-free, and reliable in a range of landscape conditions.

These plant portraits were written for a broad audience of gardeners and growers, hobbyists and homeowners, and horticulturists, professional and otherwise, of all shapes and sizes. As a result, the language is not highly technical, but the critical information you'll need is there to successfully grow, and garden with, these plants.

I have tried to share my sense of the diverse beauty and irreplaceable magic that these woody plants bring to the world. I hope this book will entice you to grow them—it is, after all, a book for people who love plants.

Introduction

by J. C. Raulston

"In any given region of the United States, 40 shrubs and trees make up 90%-plus of the landscape plantings."

RAULSTON'S "SECOND LAW OF LANDSCAPE PLANT DIVERSITY"

"The purpose of the NCSU Arboretum is to enrich and expand urban and residential landscapes by promoting a greater diversity of superior and better adapted landscape plants."

MISSION STATEMENT OF THE NCSU ARBORETUM

A Passionate Mission

The above two statements reflect the poles: The kind of plant reality that exists, and the dreams for unlimited potential that have been at the core of twenty years of passionate work at the North Carolina State University (NCSU) Arboretum. The wealth of new plant materials being introduced to horticulture each year means there is no excuse for clinging to the "old standards" out of ignorance or fear. Our mission at the NCSU Arboretum is to select, distribute, and promote high-quality woody plants that deserve a chance in our gardens and landscapes. Since 1976 we've tried over 9,000 plants from all over the world and kept careful records on their hardiness and ornamental value.

This book is another important part of our overall effort to spread the word about outstanding new ornamental trees and shrubs that are available to American gardeners. All the plants included in this book have been grown successfully at the arboretum. The chapters following grew from detailed profiles on our trees and shrubs written by Dr. Kim Tripp when she was a post-doctoral associate at the Arboretum. Most of the photographs are mine.

While all the plants you will read about have grown comfortably in North Carolina, this should really be seen as a national, even internationally focused book. The plants we've received and grown have come from some 55 countries at a rate of about 400–900 accessions a year. Horticulturists from around the country will find much to whet their appetite for potential trials in these seductive pages. Indeed, horticulturists from around the country have already provided plants for trial that have been exciting and vigorous at their varied locations, and, in friendship, these horticulturists have also acted as remote testing stations to evaluate many of our plants in the huge diversity of conditions which exist throughout America. When we wondered about the potential adaptability for an obscure but magnificent tree that showed promise in our trials, *Sinojackia rehderiana*, plantsmen friends near Houston and in central Pennsylvania planted trial plants for us and quickly demonstrated proven heat and cold tolerance as well as verifying the beauty of the plant.

The Education of a Gardener

Gardening has been a passion of mine since childhood on a wheat farm on the plains of north-central Oklahoma. From the beginning, and especially from the reality of that place, in that time, it was apparent to me that finding "different" plants and then growing them successfully are the mainstays of the passionate gardener's life. With more years of experience, endless travel, and continued exposure I now see it as a universal reality of gardening.

In the 1950s, the world of American horticulture seemed to me to exist either in England information adaptations, or in the Northeast, and specifically in Boston. I could subscribe to *Horticulture* magazine, and could read of Wyman's observations on trees and shrubs at the Arnold Arboretum—but their discussions of boxwoods, azaleas, and golden-chain trees did not seem to relate to my world and experience—or to much of the world beyond the Hudson. The realities of harsh winds, rapid and extreme temperature shifts, and low rainfall of the Plains were not to be found in their experience and recommendations. And later in life, when I moved to and gardened in what some would view as "more favorable" climates, I learned it is never truly paradise "over there." The view is expressed by Henry Mitchell with absolute perfection in *The Essential Earthman*: "It is not nice to garden anywhere. Everywhere there are violent winds, startling once-per-five centuries floods, unprecedented droughts, record-setting freezes, abusive and blasting heats never known before. There is no place, no garden, where these terrible things do not drive gardeners mad."

I quickly learned that the printed word was only a rough starting point in achieving success. Dr. Robert Ealy, a renowned teacher and major inspiration of mine at Oklahoma State University (later moving to Kansas State University), was that rarest of gardening composites with a Masters of Landscape Architecture degree for the design awareness, and a second Doctor of Philosophy degree in plant physiology for the understanding of plant adaptation and responses. Brilliant garden design requires healthy and attractive living plants to achieve the intent. An easy and obvious concept—but in reality rarely achieved or mastered. Dr. Ealy opened my eyes to what could be possible in Oklahoma (and by implication anywhere else) with an understanding of microclimates and by testing materials for wider potential adaptability. In his personal home garden, amazing plants from azaleas to desert willows inspired me to a life of constantly wanting to know, What will grow where? Why? And how can I find it to try?

I was taught, and continue to believe, that only by observing the behavior of a given plant in a given location could one determine its actual performance in the wide diversity of climates and habitats that encompass the huge and varied American continent. Each garden is an experiment station, and each gardener becomes a repository of unique knowledge about plant adaptation.

After a varied and irregular path of schools, jobs, and widely diverse horticultural experiences, I at long last found myself in Raleigh, with responsibility of teaching ornamental horticulture and working with the state nursery and landscape industries. This new job finally offered the opportunity to return to my life-long desire for focused landscape plant observation and evaluation.

The North Carolina State University Arboretum

The NCSU Arboretum, which ultimately changed my own gardening experience and plant knowledge so profoundly, was conceived in 1976 as a teaching, research, and extension facility of the NCSU Department of Horticultural Science with a mission to determine the adaptability of new and uncommon landscape plants for use in the southeastern United States and to formally promote the production and utilization of superior adapted plants discovered in this evaluation. The southeastern United States has a rich historic garden tradition, and an exceptionally varied native flora which has long yielded superior plants admired for their landscape beauty around the world. The southeastern climate is also amenable to successful cultivation of a huge diversity of exotic plants. Yet historically there have been few institutions in USDA hardiness zones 7–8 devoted to the testing and promotion of new trees in comparison to other regions of the country—and certainly nothing compared to the many distinguished institutions of the USDA hardiness zone 5–6 "arboretum belt" extending from the Arnold Arboretum near Boston to the Morton Arboretum near Chicago and on to the Missouri Botanical Garden in St. Louis.

Planting for the Future

Development of our program began in 1976 on an 8-acre site at a NCSU Department of Horticultural Science research farm, about one mile from the main NCSU campus in Raleigh, North Carolina. Planting beds were laid out in a picturesque pattern of irregular paths and island beds according to a master plan created by Master of Landscape Architecture graduate student Fielding Scarborough. This early development proceeded at a pace of about one acre a year. With essentially no budget for plant purchases, early plantings consisted of small donated plants from generous plantsmen friends through personal contacts, and plants propagated from cuttings gathered in personal travels throughout the country.

We were so eager to get something in the ground to "make it look like a garden" that plants were planted far too small and it was several years before the "twig arboretum" remained visible in winter following our rare snow accumulations. Luckily, a residue of a few larger trees and shrubs from an older attempt at a plant testing program on this site in the 1950s gave some background and illusion of gardens. I think few people at that time, probably including even myself, fully imagined the garden that would eventually grow from the tiny plantings. Even fewer could foresee the huge impacts it would have on the availability of plants to home gardeners in less than 20 years.

Among some of the early plantings that were to be significant in following years was the "dwarf conifer" (some of which are now 20 feet tall!) collection, which successfully demonstrated that conifers could indeed grow happily and well in the South, eventually spurring interest by growers and landscapers to add these superb and eminently useful plants to their programs. I remember the great excitement of the first showy, fragrant flowering one January of the magnificent Japanese flowering apricot, *Prunus mume*, finally bringing into focus for me why the Japanese were so intensely passionate about this remarkable plant grown for so long in the Orient. A display collection of ornamental grasses foreshadowed the coming flood of these plants which were emerging from German and Dutch gardens and would transform the American garden landscape in the '80s through the efforts of van Sweden and the "new American garden." In our lathhouse on raised beds of well-drained bark, we began to see the importance of root system conditions that was to become a focused part of our technical research a decade later. We learned that even the many temperamental ericaceous plants, from *Enkianthus campanulatus* (and the rare and remarkable *E. serrulatus*), *Leucothoe*, *Pieris*, and *Rhododendron* to the even less likely *Callunas* and *Ericas*, could thrive in this hot, humid climate if roots were provided a well-aerated home. Even more remarkable, the "impossible-to-grow-in-the-South" *Cornus canadensis* was grown and flowered successfully.

The endless list of over 10,000 rare species and obscure cultivars in *Hillier's Manual of Trees and Shrubs* from England, which I had read through for so many years as a virtual horticultural pornography of dreamed-of fantasies, was now an inspiration to use as a checklist to search out, collect, plant, evaluate for survival, and determine ornamental potential.

A sabbatic leave of a year on the West Coast in 1981, centered in San Francisco, brought large numbers of exotic and wonderful Mediterranean climate plants from Australia, New Zealand, southern Europe, and California back to Raleigh for trial. They had two mild years to begin to become established when the record 200-year historic low-temperature winter dampened enthusiasm for this beguiling group of plants. Visits to European nurseries, a collecting expedition to Korea, botanic garden seed from around the world, avid plantsmen friends sharing treasures from their gardens— all contributed to a river of plants which began to flow to the garden for trial.

During the first 15 years, the Arboretum development program intensely focused on plant collections rather than the development of buildings, pavings, and landscape hardscape features, a reversal of a far more common plan in most current United States public garden programs. The plants total over 9,000 received to date. They have included a diverse mixture of native plants from wild populations, exotic introductions of both wild species and horticultural cultivars, new introductions from the commercial nursery industry, and "local" cultivars of selections made by amateur plantsmen across the country. Today, the NCSU Arboretum fully occupies the 8-acre site and contains one of the most diverse collections of woody plants in the United States, with over 5,000 different species and cultivars in over 450 genera.

Practical Wisdom

As every weather-battered and battle-worn gardener knows, there is a long leap of experience from collecting large numbers of fascinating plants to showing off successful beautiful plants. And the toll of those noble chlorophyll warriors who have fallen in the climatic warfare battle in our trials is monumental indeed. One, more traditional, horticultural faculty member of our program once remarked "what a terrible horticulturist Raulston is," looking only at the withered stems and blackened stumps of failed plants that often seem so plentiful in the Arboretum at various times. But we deliberately stretch the physiological limits in every direction with constant trials of plants with likely marginal possibilities.

Our underlying philosophical view is best expressed by the noted plantsman, Sir Peter Smithers, "I consider every plant hardy until I have killed it myself." When our accession lists are reviewed, at times I have felt I've accumulated a perverse version of the birder's "life list"—a "killed life list" and a record of having destroyed a greater diversity of trees and shrubs than anyone else of my era (an achievement of dubious merit perhaps). But this long and painful process has revealed fascinating survivors.

Although plantsmen always talk incessantly about the problems of record low winter temperatures creating dramatic loss of plants, in our area we have learned the single most important adaptation factor is the ability of plants to tolerate heavy summer rains. At this time of year, temperatures are highest and active roots are requiring oxygen at their greatest level to function normally. When intense rains flood our heavy clay soils and create temporary anaerobic conditions, aeration-sensitive roots may die in a matter of hours. Firs (*Abies* spp.) are particularly sensitive to this problem, and plants have grown for several years with apparent success—and then an August afternoon with 3 inches of rain and temperatures in the 90s will occur. Bleached brown foliage and shriveled branches are evident within days and the plants die quickly. Various individual plants under evaluation have also failed from a distressingly long list of other climatic intolerances.

But the miracle of biology, and the frequent surprise, joy, and astonishment of gardeners, is that organisms do also demonstrate phenomenal tenacity and durability. The large number of plants that have survived all the various environmental stresses of our area remain for selection, distribution, and promotion if they are of significant uniqueness and/or ornamental value.

Introduction Programs at the Arboretum

Many varied types of "introduction" programs are utilized by the NCSU Arboretum to maximize the potential for successful market promotion. Our primary goal is to have recommended plants available for commercial purchase by home gardeners—at least somewhere, in some form. In contrast to many other large and well-known plant evaluation and introduction programs which focus only on commercial potential for major mass-market items, in our program a "successful market introduction" may be as small as a specialty mail-order firm producing 50 of a rare plant a year for the entire United States. Of course we are delighted if a corporate nursery produces a half-million plants a year of a more heavily desired plant for more widespread use. Both extremes and a full spectrum of the various intermediate possibilities have resulted from our work many times.

Potentially significant plants for new introductions come from origins of diverse backgrounds, history, and character. Individual gardeners may have variants in their gardens or know of local selections which may be publicized and distributed; e.g., *Gordonia lasianthus* 'Variegata', a seedling variant from a Florida nurseryman; *Bignonia capreolata* 'Tangerine Beauty', an old but forgotten introduction received from a Texas plantsman/nurseryman who had preserved it in his garden; or a "hardy" *Abutilon* first discovered growing in a private garden in Long Island, New York, and traded from garden to garden across the country until reaching us. "New" or "unused" botanical species may come from new discoveries or reintroduction of once collected and later forgotten plants; e.g., *Cornus kousa* var. *angustata* and *Loropetalum chinensis* var. *rubrum* from China; *Cercis canadensis* ssp. *mexicana* and *Cornus florida* var. *pringlei* from Mexico.

An ever-increasing list of plants have been named and promoted by the Arboretum for unique characteristics seen in seedling variants in the nursery, or which were found as mutations and then formally named for introduction. These plant cultivars include:

Bignonia capreolata 'Tangerine Beauty', *Camellia* 'Carolina Moonmist', *Cercis yunnanensis* 'Celestial Plum', *Cornus kousa* 'Little Beauty', *Cornus mas* 'Spring Glow', *Ilex* 'Carolina Cardinal', *Ilex* 'Carolina Sentinel', *Illicium anisatum* 'Pink Stars', *Illicium mexicana* 'Aztec Fire', *Jasminum beesianum* 'Goldsport', *Juniperus horizontalis* 'Lime Glow', *Koelreuteria paniculata* 'Beachmaster', *Lagerstroemia fauriei* 'Fantasy', *Lagerstroemia fauriei* 'Townhouse', *Melia azaderach* 'Jade Snowflake', *Nyssa sylvatica* 'Dirr's Selection', *Photinia villosa* 'Village Shade', *Pinus taeda* 'Nana', *Quercus phillyreoides* 'Emerald Sentinel', *Raphiolepis umbellata* 'Blueberry Muffin', *Styrax japonica* 'Crystal', *Styrax japonica* 'Emerald Pagoda', *Ternstroemia gymnanthera* 'Burnished Gold', *Ulmus alata* 'Lace Parasol', and *Viburnum awabuki* 'Chindo'.

The Advocacy Program. Perhaps the most important of the programs in total impact is a wildly diverse "advocacy" program, which begins by promoting promising plants through widely varied printed professional media such as *American Nurseryman*, *Nursery Business*, the *North Carolina Association of Nurserymen Nursery Notes*, *Southern Nurserymen Association's Proceedings of the Research Workers Conference*, journals of assorted specialty plant societies, and so on.

A unique and an especially valuable outreach to the public gardening world in North Carolina was a three-year program of weekly university press releases on selected arboretum plants of seasonal interest throughout the year. These were written by Dr. Kim Tripp, then an Arboretum post-doctoral associate. Dr. Tripp has the rare combination of a keen observational eye of a technically trained scientist, the soul and vision of a poet to see the rarely recognized inner beauty and secrets of these plants, and the technical proficiency of a practiced professional writer to clearly convey this knowledge and awareness. Her articles were distributed to newspapers and horticultural journals throughout the region and were widely used and reprinted. In an attempt to make these exquisite portraits available to a larger deserving audience, this long series of highly informative and beautifully written plant profiles eventually became the foundation of this book.

Outreach communication also continues at the Arboretum, with lectures presented each year to over 10,000 people in about 20 states to make both nursery producers and gardening consumers aware of the new and improved plants that are entering the American commercial market and to recommend further production or landscape trials.

Although the Arboretum does not have a formal breeding and cultivar introduction program, it has significantly impacted the range of plants being grown in the nursery industry today through its relatively unique broad-based advocacy program. One successful strategy has been a program of plant distributions for grower trials, which began in 1980. This has steadily grown in size and impact to the point that over 10,000 plants are now propagated and shared annually. Getting plants in the hands of growers where they can live with them daily and see them at their moments of special glory can go far beyond normal dry, written promotional blurbs in convincing them to add these special plants to their standard production lines. This popular program is in its 15th year, with about 65,000 individual plants formally distributed of about 320 different plants for trial to date. A few examples of the diversity of plants in current commercial nursery production which have their origin in this type of distribution would include: *Abelia chinensis*, a deciduous flowering shrub and outstanding butterfly and fragrance plant; *Cercis canadensis* ssp. *mexicana*, a small flowering tree from Mexico with beautiful small, undulate, glossy foliage; *Morus australis* 'Unryu', a deciduous tree with dramatic contorted branches originating in China long ago; and *Vitex rotundi-*

folia, a salt-tolerant deciduous groundcover shrub collected in Korea for use in beach areas. It is estimated that at least 250 different plants are currently in commercial production in the United States directly as a result of the NCSU Arboretum distribution program.

The arboretum also develops information and plant display booths for nursery short courses. Educational displays are made which promote promising plants. Various recent shows have featured a collection of some 20 or more different varieties of such nationally significant specialty collections of the Arboretum, such as *Cryptomeria japonica*, *Ilex*, *Mahonia*, and *Nandina domestica*. Such visual promotion, along with written information and access to cuttings from the Arboretum, helps stimulate commercial production on a larger and more widespread scale.

New crop propagation workshops are conducted with hands-on sessions at the Arboretum. Commercial growers are taught propagation techniques, shown new recommended plants growing in the Arboretum, and are allowed to collect propagation wood while at the program—or are encouraged to come back at appropriate times to do so. This unique open invitation for commercial nurserymen to come to the Arboretum to collect propagation wood of plants of specific interest to them, under the guidance and supervision of the Arboretum staff, is probably the most significant distribution program of the Arboretum in its total effect on the industry. Growers from North Carolina and from throughout the United States are in the Arboretum every week throughout the year. It is estimated that about 400,000 cuttings are collected annually at present, and that over 3 million cuttings have gone out through this program over the years. On a more limited basis, propagation materials are also shipped across the country and around the world upon request from those commercial growers unable to visit the Arboretum.

Criteria for Marketability

One of the hardest and most painful lessons we've had to learn has been that an environmental-stress tolerant plant with great ornamental beauty and appeal is not necessarily a guarantee of commercial success, and that such beauty may have no relationship to public acceptability and use. In general, for widespread commercial success, a plant must have a good balance of at least several of four production and utilization factors.

1. The grower must be able to propagate the plant efficiently (parent stock plants or seed stock available; good success rate).

2. The plant must be economically "efficient" in production (few biological problems, low losses and culls, and above all else—grow fast).

3. The plant must be a tough survivor in the landscape once planted (stress tolerant, able to survive cultural abuses by both retailers and unknowledgeable gardeners).

4. The plant must be either an already familiar plant (new forms of familiar plants like azaleas, dogwoods, lilacs, etc.) or be colorfully showy and appealing in the five-week peak spring sales period when frenzied gardeners are released from winter madness and require planting and color therapy for survival (and when 70% of all plant sales occur).

Many of our finest plants in an ornamental sense have hit these barriers and thus remain either unavailable or in very limited availability through focused specialists on a rarity collector basis. The Japanese flowering apricot (*Prunus mume*), for example, has had probably more promotion than any other single plant from our program, but after 15 years of persistent work, it is still very limited in market availability today—due entirely to the marketing reality that gardeners are not shopping in nurseries in January and February when it blooms. Later, when gardeners are in the April buying frenzy, they are distracted and mesmerized by the many spring blockbusters like azaleas, cherries, and dogwoods, while the *Prunus mume*s just sit there in naked embarrassment, two months past their previous ignored glory.

Other exceptionally fine ornamental plants that we've promoted strongly for years that were left at the horticultural altar include such things as a genetic dwarf loblolly pine (extreme difficulty in grafting propagation); *Juniperus deppeana* 'McFetters' (little stock plant wood, low percentage cutting take, and slow nursery growth); *Magnolia grandiflora* 'Hasse' (difficult to root or graft); *Euscaphis japonica* (limited seed availability and difficult germination); *Heptacodium miconioides* (showy in fall with limited buyer interest in October shopping); *Cercis canadensis* 'Oklahoma' (budding very difficult); and *Styrax japonica* 'Emerald Pagoda' (cuttings root but die over winter).

By contrast, in the same era, Leyland cypress, ×*Cupressocyparis leylandii*, has become a major commercial plant in spite of no showy flower, fruit, or foliage color, no seasonal interest, little promotion, and potential bagworm and canker problems. Why? Because it is relatively successful in propagation and ample cuttings are available; because it has potential for 3–4 feet of annual growth (in our area); because it grows quickly to market size, providing both good profit margin for the grower and good perceived market value to the consumer; because it fits a widespread and specific market use of screening of smaller properties from neighbors in crowded neighborhood; and because it requires little care once planted.

Inspiration for Gardeners

Our desire is to provide inspiration and ideas from an American experience base for American gardeners to try an expanded range of exciting trees and shrubs beyond those 40 standards presently available in their community. This book presents only a tiny sampling of the possibilities that exist with an estimated number of over 15,000 different trees and shrubs available at some level somewhere in the United States marketplace today. Almost any plant one could possibly want is available today if one wants it badly enough to hunt for it with dedication and passion.

The bold gardener by necessity will become an independent explorer and an adventurer, as good reference materials for cultural information and plant performance often takes years to develop because of the time required for a new tree cultivar to mature, show all its characteristics, and reveal its idiosyncracies. Public gardens are increasingly focused on beginning gardener orientation to standard gardening techniques, and fewer gardens today assemble large collections of new woody plant cultivars for evaluation and viewing by the public. In many areas of the United States, there is a limited presence of public gardens of any kind to refer to. Specialized plant societies bring like-minded collectors together for information and plant networking and exchange through formal newsletters, journals, annual meetings, seed exchanges, and so on, and can be one of the best sources of detailed, if not pointedly focused plant information.

But there is also an exciting and wonderful network of highly individual, passionate (some say crazed) gardeners located all across America in the most unlikely places who are eager to share information about their plant trials, and in many cases plant starts as well. I've seen the most phenomenal collections of plants in private gardens in little towns in Vermont, Kansas, Texas, North Dakota, Arkansas, etc., often containing likely the only plants of their kind in the state, and very often plants "that simply can't be grown in this climate"—southern magnolias in Toronto, palms in Boston, and spruces and firs in Mobile. My fondest "location expertise" memory is of a slide lecture I gave long ago in Denver (when I was a little younger, a lot smugger, and frankly probably showing off a bit), with a fairly esoteric listing of obscure trees for potential trials in the Intermountain area of the West. After the talk, a quiet tall, lanky cowboy in hat and boots came up to talk to me and allowed that "that's a right interestin' list you got there, young fella. I grow most of them and these are the ones that do the best"— as he went off into a detailed, knowledgeable discussion of his treasures. He was from the northern Utah region near Winnemucca—not where one would have ever expected such expertise, nor such remarkable success. I have another garden friend studying zero maintenance landscape plants for eastern Oregon, at his summer home in an area of less than 10 inches of rain a year with bitter winter lows and searing summer heat— and he's finding treasures that work. Such exciting pioneers can be found in every section of this country.

In the final analysis, one of the most exciting parts of this professed philosophy of experimental gardening is that it allows individuals to truly be individuals. It allows you to grow plants in your area that no one has ever grown before, and to know information of adaptation that no one else knows or that is not written down in "reference books." In our crowded world of increasingly standardized products and culture, few such opportunities exist to be so strongly individual and experimental.

Another of my endless laws of landscape plants states that "You can throw a dart at a map of any location in the United States and within a half-mile of that impact point, you can find totally new and wonderful plants that have never been seen before that have new ornamental interest and potential." These plants arise from seedling variations within both native plants in the woods and on the roadsides and exotic plants in the landscape, and from mutations of growing tissues on plant branches. One only needs to learn how to truly "see" the plants around one every day. With a little experience, you will find remarkable unknown plants are everywhere you look.

I hope this book will share some of the joy and excitement of the magic and beauty possible from trees and shrubs at all times of the year, in the collections of the NCSU Arboretum and elsewhere. I hope it inspires and encourages you to follow our program and plant in your own garden the releases that are constantly coming from our introductions.

Spring

Aesculus spp. / Buckeyes and horse chestnuts PLATES 1, 2

Buckeyes and horse chestnuts are members of the genus *Aesculus*, a large group of many gardenworthy plants that range in size from low and shrubby to tall and imposing. The genus includes North American natives, European and Indian exotics, and many horticultural hybrids. They produce beautiful candelabras of flowers among coarse, leafy branches with hand-shaped foliage, and bloom in May through June and July (exact time depends on the species or cultivar).

All of the trees discussed here are adaptable to most landscape conditions, except for extremely dry soils. Cold hardiness varies, depending on species and cultivar. *Aesculus* prosper in full sun or partial shade, with full sun giving the best flowering and also minimizing fungal disease problems on the foliage. They are sexually propagated by seed layered in a moist medium, such as peat moss, and stored for three to four months at 40°F (4°C). Cultivars, with the exception of *A. parviflora* selections, are generally grafted onto seedling understock.

Among the best known of the group is the common horse chestnut, *Aesculus hippocastanum*, a native of Greece and adjacent mountainous areas. This large, bold-textured tree will reach 50–80 feet (15.2–24.8 m) in height with an almost equal spread of its beehive-shaped crown. Creamy flowers with a yellow throat and reddish speckled center are borne in 6–8-inch (15.2–20.3-cm) pyramidal clusters in June. Flowers mature to fruits that are 1–2 inches (2.5–5 cm) in diameter, held in a shell covered with dangerously sharp spines. The shell splits open to reveal the appealing chestnut-brown, shiny nuts that are the treasure of children the world over.

The common horse chestnut is an excellent tree for large public spaces and has been widely planted in parks and recreational areas in both the eastern United States and Europe. It is not a good choice for small properties (there are many good, small *Aesculus* for those settings), but it is very useful and striking in large, rolling pastures or other open lawn areas. Fall color is essentially nonexistent except for perhaps some gold-yellow. The foliage is susceptible to powdery mildew and leaf blotch and often browns and dies early in the fall. The excellent cultivar 'Baumanii' has large, double white flowers that are sterile and therefore produce no fruit (which means no spiny fruit to rake or step on). *Aesculus hippocastanum* is one of the more cold-hardy of the *Aesculus*, with reliable performance through zone 3.

Among the numerous North American native *Aesculus* species, there are five native to the eastern United States (with somewhat overlapping ranges) that are good garden plants or are parents of the better horticultural hybrids. They range from small to large trees, with beautiful flowers. These species can be found in the woods from Georgia through the Carolinas and up into Ohio. Each of the five species has its own geographic area of concentration, but they do overlap and are easily confused unless they are in flower (and even then they are difficult).

Aesculus flava, yellow buckeye, is a large tree, ultimately reaching 50–70 feet (15.2–21.7 m) in height. Yellow flowers appear in May and are quite showy in 6-inch (15.2-cm) upright clusters. The fruit is smooth-skinned and eye-catching because of its irregular, globular shape. The foliage, which may turn burnt orange in the fall, is less susceptible to fungal disease than that of other species. Yellow buckeye has a northerly native range, from areas of northern Georgia to the mountains of North Carolina and Tennessee, up through Pennsylvania, Ohio, and Illinois. It is reliably hardy through zone 3.

Aesculus glabra, Ohio buckeye or fetid buckeye, is another native, with its range stretching from Alabama north through Tennessee and up through Ohio. It can be distinguished from its other native cousins by a fetid odor released when the stems are bruised or broken. Flowers are greenish yellow with orange throats and not as showy as other species, but the blooms are delightful additions to the late spring woodland landscape. The gray bark is often corky. The foliage is a very attractive rich green and emerges relatively early. Ohio buckeye is more frequently seen as a young tree in native settings, but it will reach 20–40 feet (6.2–12.4 m) in height or more in most landscape settings, with a pleasing rounded habit. *Aesculus glabra* is reliably hardy through zone 3.

Aesculus parviflora, bottlebrush buckeye, is a spreading, multistemmed, low, shrubby plant reaching only 6–10 feet (1.8–3.1 m) in height (Plate 1). The form is fluid and beautiful. Its native range is from South Carolina through Alabama and Florida. The moderately pubescent (fuzzy) foliage, rarely disfigured by disease, often turns a lovely clear yellow in the fall. Flowers are white and appear late in June on exceptionally large, 8-inch (20.3-cm), erect clusters. Fruit is smooth and light brown, but fruit set is often sparse.

Bottlebrush buckeye is extremely tolerant of shade and an excellent choice for small, shady gardens where its white spikes of flowers make a unique summer show. It is reliably hardy as far north as zone 4. It is important to note that it will sucker and spread. In very restricted areas, the plant may need root pruning, just outside the dripline, each year in the spring. It can be propagated from softwood cuttings rooted under mist. The botanical variety *A. parviflora* var. *serotina* flowers several weeks later than the species and has smooth foliage. The cultivar 'Rogers' is a selection of this variety, from the University of Illinois, with incredibly grand, 18-inch (45.6-cm) flower clusters that bloom very late. 'Rogers' is less likely to sucker.

Aesculus pavia, red buckeye, is another low-growing, shrubby species that forms clumps reaching 10–20 feet (3.1–6.2 ft) in height with equal or greater spread. This species is found from the coastal plain of North Carolina south to Florida and west all the way to Texas. Flowers are generally a very beautiful salmon to cherry-red color, but there is a yellow-flowered native strain as well, *A. pavia* var. *flavescens*. Foliage is attractive in spring and summer but develops no appreciable fall color. Red buckeye usually has less powdery mildew than horse chestnut but may have severe leaf-blotch problems late in the summer. The cultivar 'Atrosanguinea' has very deep red flowers. 'Humilis' is an especially low-growing, almost prostrate form with small clusters of true red blooms. Red buckeye is cold-hardy through zone 4.

There is some argument among botanists and horticulturists as to whether *Aesculus splendens* is truly a separate species or only a botanical form of *A. pavia*. It is also a red-flowered form, with its chief differentiating characteristic a very heavy pubescence on the undersides of leaves. It is less cold-hardy than *A. pavia*, being reliable only through zone 6.

Aesculus sylvatica (previously named *A. georgiana*), painted buckeye, is abundant in the Piedmont woods of North Carolina, Georgia, Tennessee, and Alabama, with some of the population in Virginia. Flowers range from yellow to yellow-green to pinkish yellow and are borne in broad clusters, 5–6 inches (12.7–15.2 cm) high, with color variation often on the same plant. This understory tree is generally 5–12 feet (1.5–3.7 m) tall with a somewhat rounded, open habit. Foliage is an attractive emerald green but shows little fall color. It is reliably hardy through zone 6.

Some of the best *Aesculus* for the garden are the horticultural hybrids. Red horse chestnut, *Aesculus* × *carnea*, is a spectacular medium-sized tree reaching 20–40 feet (6.2–12.4 m) in height with a rounded to broadly conical shape. The glossy foliage is a beautiful grass-green and makes an excellent foil for the 8-inch (20.3-cm), upright candelabras of salmon-pink to light red flowers. Disease is less of a problem than with common horse chestnut, but powdery mildew is nonetheless likely to appear. This does not interfere with the overall growth or flowering of the tree, but results in early leaf drop. The cultivar 'Briottii' (Plate 2) is a floriferous salmon-flowered form, while 'O'Neill' offers truer red flowers in particularly large, 10-inch (25-cm) clusters, and 'Rosea' has pink-red flowers. This hybrid is hardy to zone 4.

Aesculus × *hybrida* is something of a catchall name for the many hybrids of *A. flava* with *A. pavia*. Flowers are yellow-red to salmon-red to pink. The true names of these hybrids are very confused and they are often mixed with other hybrids having *A. pavia* as one of the parents. 'Autumn Splendor' is a cross of *A.* × *hybrida* with *A.* × *glabra*, from the Minnesota Landscape Arboretum, which offers the unusual trait of bright red fall color.

The horse chestnuts and buckeyes are a large and often confusing clan, but the selections and hybrids found in reputable nurseries and garden centers offer unusually bold-textured foliage combined with lovely floral displays for late spring and summer. This unique combination of landscape character makes *Aesculus* a marvelous choice for most gardens. Whether you choose a low-growing beauty for close inspection or a magnificent large tree to display those arresting flowers, *Aesculus* will bring a strong show of color and texture to your garden.

Akebia quinata / Five-leaf akebia   PLATES 3, 4

Five-leaf akebia is most often noticed in the early fall when its fat, interesting fruit hang from the vine, but it is also a marvelous plant throughout the entire growing season. Deciduous to semievergreen, this twining, climbing vine displays divided foliage, with each leaf consisting of five individual leaflets grouped into five "fingers" of an oddly hand-shaped leaf. The foliage of five-leaf akebia is a beautiful, soft blue-green during the spring and summer, but there is no fall color display.

Each spring, the vine is covered with dark wine-purple, 1-inch (2.5-cm) flowers. The chambered, deeply colored blooms exude a light, fruity scent that is a welcome harbinger of the heady aromas of summer in the garden. The flowers are at their peak when the new foliage has already emerged and so require reasonably close inspection for full appreciation. Looking up into the flowers dangling from an arbor or trellis is the perfect way to see them. Male and female flowers are different from each other, with male flowers often being lighter in color. Female flowers mature into fabulous 6–8-

inch (15.2–20.3-cm) fruit that hang from the vine in fall like dozens of purple bratwurst. The pulp is actually edible. The fruits split open to reveal rows of shiny black seeds as they mature.

Five-leaf akebia is an incredibly tough native of China, Korea, and Japan that will adapt to almost any site in sun or shade, or wet or dry conditions. It is hardy through zone 4 and is readily propagated from cuttings taken in spring and early summer and rooted under mist.

Akebia is a very rapidly growing vine and is an excellent choice for providing three-season interest quickly on almost any landscape surface. New fences, old neighborhood boundaries, trellises painted just the wrong shade, and budget-lumber arbors can all be effectively covered with this delightful vine in almost no time. It is pest- and disease-free and requires little maintenance except, perhaps, periodic pruning to keep it in bounds.

A few cultivars of *Akebia quinata* that can be found with a bit of persistent hunting in specialty nurseries. 'Alba' and 'Shirobana' are white-flowered and white-fruited forms with lighter green leaves and slightly less vigorous growth rates than their darker cousin. 'Alba' is slightly more evergreen in nature and blooms a bit earlier with more fragrance. 'Rosea' has light magenta or lavender flowers, almost as if it had faded to a more delicate tint than the species, and 'Variegata' has white-variegated foliage. Two other species of *Akebia* may also be available. *Akebia trifoliata*, three-leaflet akebia, is also a China native. Its leaves are divided into threes instead of fives, but otherwise it is similar to *A. quinata*. It is a less vigorous climber and may be slightly less hardy than *A. quinata*, but it has the same dark red flowers and fruit. *Akebia* × *pentaphylla* is a hybrid of *A. quinata* and *A. trifoliata* with traits intermediate to those of its parents.

The intricate carpet of divided foliage and deeply colored flowers makes an inviting cover to an entry walk. At the North Carolina State University Arboretum, a number of different *Akebia* species and cultivars weave delightfully around each other on the arbor at the Arboretum's entrance. Let five-leaf akebia's tracery of inviting greenery and fragrant flowers welcome you into your own garden throughout the season.

Amelanchier spp. / Serviceberry   PLATE 5

Serviceberry trees bring refined interest to the garden in all four seasons of the year. These ornamental small trees can be any one of a number of species of *Amelanchier*, a group of deciduous flowering shrubs and trees native to various parts of North America. A number of species are rather similar in appearance, which results in frequent botanical confusion over their correct names. In recent years, a large number of hybrid cultivars have become available, which complicates the name issue even further, but whether native species or hybrid cultivar, all serviceberries are gardenworthy. Dr. Tom Ranney, of North Carolina State University's Mountain Horticultural Crops Research Station in Fletcher, North Carolina, has done extensive work with these trees to illustrate their landscape potential.

All *Amelanchier* species and hybrids bloom in the spring, producing clusters of delicate white flowers. Their smooth gray bark is the perfect foil for the creamy white blooms that gleam like freshwater pearls on gray velvet. In the fall, the leaves are brightly colored, and in winter, the tree's form and bark are quite elegant. Serviceberries are known for their sweet, huckleberry-like fruit, prized by both gardeners and birds (which generally beat the gardeners to the tasty harvest). The fruits turn red in summer and mature to a dark gray-black when ripe. All serviceberries are completely hardy through zone 4. They will tolerate a range of landscape conditions from heavy, wet clays to dry, poor, sandy soils. Seedlings may be subject to fireblight, leaf spot, and rust problems, but most cultivars are essentially pest- and disease-free.

Amelanchier arborea, downy serviceberry or shadbush, is native to much of the eastern United States, from New England to the southeastern Piedmont. It is the most tree-like of the serviceberries (hence, "arborea") usually possessing one main trunk and reaching heights of 10–25 feet (3.1–7.5 m). The leaves emerge a soft, downy gray and gradually turn dark green for the summer. In autumn, the foliage of downy serviceberry has outstanding yellow-gold or orange-red color. The fruits of downy serviceberry are among the largest and best tasting of the amelanchiers. The fruit makes tasty pies and confections that taste similar to but slightly more tart than blueberries.

Amelanchier canadensis, the shadblow serviceberry, is also native throughout the eastern part of the United States, and southeastern Canada. It is easily, and often, confused with *A. arborea*, to such a point that much of the nursery stock called *A. canadensis* is actually *A. arborea*. Shadblow serviceberry is a lower, more shrubby plant than *A. arborea*, multistemmed and reaching heights of 5–15 feet (1.5–4.5 m), and with a somewhat later bloom time. In the wild, it is found in bogs and wet sites, while *A. arborea* tends to grow in drier waste areas and better-drained wooded sites. The chief difference between the two species is the multistemmed, suckering habit of *A. canadensis*, which is lacking in *A. arborea*. The flowers, bark, and fruit of shadblow serviceberry are as attractive as those of its more arboreal cousin, downy serviceberry, but the plant's habit is a bit more ungainly.

Amelanchier laevis, Allegheny serviceberry, can also be found growing throughout the eastern half of the United States and southeastern Canada. It is somewhat intermediate in habit between downy and shadblow serviceberries, but is generally a small, distinctly treelike plant with two to three trunks. The foliage of Allegheny serviceberry sets it apart from other *Amelanchier* spp., as the new growth emerges a bronzed purple color. In years when late-spring cold delays flowering, the flowers may still be present when the leaves emerge; the combination of bronze-purple new foliage with pristine white flowers is spectacular. The leaves turn a glossy dark green for the summer and develop bright, clear yellow and red color in the fall.

Amelanchier × grandiflora, apple serviceberry, is a natural hybrid of *A. arborea* and *A. laevis*. Flowers are generally larger and many are pinkish in bud. Most (but not all) of the cultivars available in nurseries are selections of this hybrid species.

Some of the best cultivars include 'Autumn Brilliance' (from Bill Wandell in Illinois), with persistent leaves that prolong the excellent color of the fall season, and rapid growth; 'Autumn Sunset', (selected by Michael Dirr in Georgia) with "pumpkin-orange" fall color and exceptional drought tolerance; 'Ballerina', from the Netherlands, with large flowers nearly 1 inch (2.5 cm) across and purple fall color; 'Cumulus' (from Princeton Nurseries, New Jersey), a tall tree form with exceptionally abundant flowers; and the "Royal Family" series, which includes 'Princess Diana', 'Prince Charles', and 'Prince William' (the shrubbiest of the three), with abundant flowers, excellent fall color, and attractive form. 'Robin Hill' offers distinctly pink buds that fade to white flowers and good tree-like habit, and 'Rubescens' reaches 20 feet (6.2 m) with rosy pink buds that open to very pale pink flowers.

The *Amelanchier*s have been difficult to propagate vegetatively. In the past, many were grafted which led to problems with understock suckering and overgrowth of the desirable form grafted onto the rootstock. Now, with new methods of tissue culture, grafting is unusual and the overgrowth problem is rare. Procedures for rooting cuttings have also been refined so that late spring softwood cuttings can be successfully rooted.

Rooted cuttings should be left in the original rooting medium for the winter following rooting and can then be potted up the following spring to avoid loss. As with lilacs (*Syringa* spp.), the window for success with rooting cuttings is short, being a brief period in late spring. Seed can be successfully germinated (if you can snatch it from the birds) by layering it in moist peat and allowing three to four months of cold treatment.

Serviceberries are one of the treasures of the natural environment. Their cultivated forms bring these native delights to the landscape with suitably gardenesque character. The appealing flowers and bright fall color, combined with attractive form and small scale, call for wide use in landscapes.

Bignonia capreolata / Crossvine   PLATE 6

Vines are popular choices for a wide range of garden styles. Their adaptability to growing conditions, ornamental qualities, and textural character give vines the opportunity to contribute in many ways to every landscape space.

The vigorous crossvine is an outstanding example of this combination of adaptability with special beauty. Native to the eastern United States, with an indigenous range that stretches from Illinois south and east to Florida, crossvine is an excellent choice for gardens of all styles throughout much of the United States.

This semievergreen to evergreen climbing vine is covered with masses of salmon-orange to reddish brown, 2-inch-long (5 cm), trumpet-shaped flowers for three to five weeks, from late spring into summer. The flowers are variable in color among wild populations, but they generally have yellow-orange throats. Set against the dark, glossy leaves, which can reach 4 inches (10.2 cm) long, the coralline colors of crossvine's flowers are brilliant. The foliage is attractive in itself, with formal, narrowly oval-shaped and distinctly pointed leaves. In warmer areas, where this vine is completely evergreen, it adds texture to the winter garden, especially in combination with other vines of contrasting form. Its common name refers to the outline of the stem's cross-section.

Crossvine is completely hardy through zone 6. It will prosper in almost any landscape site and is free of pest and disease problems. The plant is tolerant of wet clay soils and will also grow in deep shade (although flowering will be sparse in heavy shade). Crossvine is readily propagated from seed sowed directly after harvest, or from softwood cuttings taken in early summer and rooted under mist.

This vigorous plant climbs using tendrils and "holdfasts," small circular appendages that enable the vine to cling to, and therefore climb, seemingly impossible walls, such as flat wood or porous concrete surfaces. It requires pruning and training in small gardens but is a fine choice for softening new fences, covering unattractive poles, or for bringing new interest to vertical features that have worn a bit thin in the garden.

Surprisingly, there are few cultivars or color forms. 'Atrosanguinea' is a red-flowered form with slightly narrower blooms than the species; the flowers open to an inconspicuous orange-red throat. 'Tangerine Beauty' (Plate 6), a bright, deep orange-colored form originally distributed by Wayside Gardens, is a star of the patio trellis at the North Carolina State University Arboretum. This incredibly floriferous cultivar has been undergoing propagation for renaming and re-release from Wayside Gardens and is now available in their catalog.

Crossvine brings a cheerful burst of vigor and color to the garden. Its semievergreen nature makes it valuable in all seasons, and its robust character makes it well adapted to gardens in a wide range of conditions and climates. It's a vine that's well worth consideration.

Catalpa spp. / Catalpa   PLATE 7

One of my favorite childhood memories is of the annual wait for the frilly flowers of *Catalpa speciosa*, hardy catalpa, to mature into gigantic long, thin "beans," which dangled from the tree like those on Jack's legendary beanstalk. These fruits made especially fun toys for all sorts of games—from pretend gardening to rather less idyllic "bean wars."

As I myself grew, I began to appreciate the flowers and tough nature of this plant equally as much as its appealing fruit. I was dismayed to learn, upon entering the halls of horticultural academe, that catalpa is often placed in that rather arbitrary and awful category of "weed tree." This appellation is rarely deserved by any plant unfortunate enough to be placed in this group, but it is especially inappropriate for catalpa.

Two species of catalpa are commonly seen in residential gardens and landscapes. Both are United States natives with overlapping ranges. *Catalpa speciosa*, hardy catalpa, is native to a small area of the Midwest (from Indiana to Arkansas along the Ohio River) but is naturalized in the Northeast as well. It is the larger of the two common catalpas, reaching 50–70 feet (15.2–21.7 m) in height with an open spread of up to 40 feet (12.4 m) and coarse habit. *Catalpa bignonioides*, southern catalpa, is smaller in height with a more rounded canopy but equally coarse branch structure and, as the name suggests, is more common in the southeastern states, including Florida and Louisiana.

Both species are exceptionally agreeable trees producing large, heart-shaped, grass-green leaves with a decidedly tropical appearance, and which are topped in late spring with upright, pyramidal clusters of bell-shaped white flowers. The leaves do not develop any significant fall color in the warm Southeast but can turn canary yellow in more northerly gardens. The flowers are speckled on the throat with yellow and purple and are somewhat similar in appearance to those of *Aesculus* spp. (the horse chestnuts). The flowers of *Catalpa bignonioides* bloom later and show more purple speckling than those of the northern species. The clusters of blooms of both species are 6–12 inches (15.2–30.4) tall and quite showy, even from a distance, as they rise up from the foliage like floral candelabra throughout the canopy. In late summer and early fall, the flowers mature into "beans" or "cigars" (catalpa is sometimes called the cigar tree), which eventually fall from the branches. Inside each fruit are a number of interesting dry, winged seeds with fringed wings.

Catalpa is an exceptionally tough landscape plant. It is well adapted to almost every modern landscape setting in full sun or part shade and will tolerate a broad range of soils and moisture conditions, from dry sand to heavy clay. *Catalpa speciosa* is cold-hardy through zone 4, while *C. bignonioides* is cold-hardy to zone 5. Catalpa is generally somewhat weak-wooded and may lose large branches during strong storms. Judicious pruning begun when the tree is young not only helps with this problem but can minimize any ungainly habit that develops over time. Some of the foliage will succumb to powdery mildew in wet years but the tree itself will continue to prosper. Catalpa is often one of the few remaining trees surviving on lots that have been surrounded by urban sprawl and construction—it will flower and fruit reliably in the face of an army of bulldozers.

There are some interesting named hybrids and selections of catalpa, and a number of Asian species worth consideration. Two cultivars of southern catalpa are 'Aurea', with canary-yellow leaves that rapidly fade to green in summer heat, and 'Nana', a rounded dwarf form that rarely flowers or fruits and is usually top-grafted onto a standard to make a pom-pom-style tree. *Catalpa bungei* is a small, shrubby tree native to China, similar in appearance to *C. bignonioides* 'Nana' but with some lobing on the leaves. *Catalpa fargesii*, Farges catalpa, is also native to China and bears distinctive pink-speckled flowers. *Catalpa ovata*, Chinese catalpa, is a third species from China with exceptionally long and thin fruit and yellow-cream flowers. *Catalpa* × *erubescens* is the classification for a number of hybrids between *C. ovata* and *C. bignonioides*, including the rich burgundy-leaved form 'Purpurescens', whose leaves and petioles emerge deep wine-purple in the spring, but fade to green in the warm South where temperatures climb rapidly to the nineties and stay there. Sadly, no doubt due to catalpa's undeserved 'weed tree' reputation, it is almost impossible to find any selections of catalpa in mainstream commercial nurseries. Gardeners will need to hunt through specialty mail-order catalogs or propagate their own plants, which is not difficult. Catalpa can be propagated readily from fresh seed or by grafting, or it can be rooted easily from softwood stem cuttings or from root cuttings taken in early winter.

Like other so-called weed trees (e.g., box elder, *Acer negundo*; tree of heaven, *Ailanthus altissima*; royal paulownia, *Paulownia tomentosa*), catalpa has many appealing traits. Its toughness, adaptability, and apparent commonness have caused it to be branded a weed, yet those characteristics are actually very valuable in facing the demands of today's urban landscapes. Explore the gardens of old neighborhoods to discover the catalpa, and see just how beautiful what is already in our own backyards can be.

Cercis canadensis / Eastern redbud   PLATES 8–11

Our native woods and fields come alive each spring with clouds of color, the flowers of myriad woody plants in bloom. In the east, we generally think most often of our native dogwood, *Cornus florida*, as chief among those flowering trees that bring the woods alive each year. But there is another small tree whose flowering display offers as much to this dramatic spring beauty as dogwood, but with less recognition—our native redbud, *Cercis canadensis*.

In spring, the woods are filled with an unforgettable, bright haze of rosy purple created by the flowers of redbud. Redbud's flowers, which bloom before the leaves emerge in early spring, are borne in an unusual manner, with the buds appearing to emerge directly from the bark, scattered up and down the branches, and on parts of the trunk. The buds are actually produced in unbranched clusters called *fascicles*, created by individual flowers bound together at their base (similar to the clustered needles of pines, which are also called fascicles).

The flower buds are a unique, bright magenta color which lightens to rosy pink as the small flowers expand to full size of 0.5-by-0.5 inches (1.2-by-1.2 cm). Flowers are borne in clusters of 4 to 10, with each individual flower on a slender, 0.5-inch-long (1.2 cm) pedicel, and with all of the pedicels of a single cluster emerging from a single node (location on the twig). Redbud flowers are fascinating to observe closely—which is a distinct advantage to having one of these trees in your own garden.

Redbud flowers are like those of peas or beans, with an irregular structure. They mature into light brown, flattened pods, 2–3 inches long (5–7.6 cm). In years with especially prolific fruit set, thousands of pods hang from the branches through the fall and into winter. Some find them unattractive, but I think the pods add an interesting element to the late-season garden, especially as they persist after leaf-fall.

Eastern redbud is a small, leguminous tree (i.e., a woody relative of beans and peas) with charcoal-colored bark and lovely, heart-shaped leaves. It reaches 20–30 feet (6.2–9.0 m) in height with a nearly equal spread and is usually multitrunked. Its many-branched, horizontally spreading crown and zig-zag pattern branching make redbud a gracefully architectural small tree for the garden.

Redbud foliage is very distinctive—like small, grass-green valentines dangling from the branches. Fall color is generally unremarkably light yellow in most years, but can be a bright canary in years when the season is especially conducive to good fall color development (bright sunny days and cool nights), especially in the northern half of North America.

Representatives of the genus *Cercis* can be found throughout most of the United States, in parts of Mexico and along a few areas of the Canadian border around the Great Lakes, as well as in areas of China, Japan, Korea (*C. chinensis, C. chingii, C. racemosa*), and the Mediterranean (*C. silaquastrum*). Some of the redbud cultivars available today are selections of the Chinese species.

Eastern redbud, *Cercis canadensis*, is native to most of the eastern half of the United States, including Texas and Oklahoma, and extends just across the border into the Great Lakes area of Canada. It is not found along the coast (it is not especially salt-tolerant). The taxonomy of the two distinctive races of eastern redbud found in Texas, Oklahoma, and Mexico are the source of great debate, with some authors separating southwestern and Mexican populations into separate species (*C. canadensis* ssp. *texensis* and *C. mexicana*), and others naming them as botanical varieties of *Cercis canadensis*. Another, undisputed species, western redbud (*C. occidentalis*), is found in California and in areas of nearby western states.

Eastern redbud is well adapted to a range of conditions. In the wild, it can be found growing from moist bottomlands to dry uplands, often on chalky soils. It is generally an understory tree in nature and is quite tolerant of moderate shade (but you will notice the heaviest flowering on the edges of woods where there is the most light). In cultivation, eastern redbud thrives in a range of soils, including heavy clays and both acid and alkaline soils, but it prefers moist, well-drained loams. It performs well in both full sun and partial shade but flowers most prolifically in full sun. A canker disease can attack and ultimately kill trees, especially older individuals near the end of the relatively short, 30- to 50-year redbud lifespan, but it is not generally a problem on unstressed, vigorously growing trees.

Eastern redbud can be hardy through zone 4, and is successfully being grown in parts of zone 3; however, hardiness can vary, depending on the source of the parent tree. Since eastern redbud is native to such a wide range of climates, it is important to be aware of the seed source of the purchased trees, or parent tree, not the location of the nursery. Trees grown from seed collected from trees native to Georgia will not be cold-hardy in Chicago or Boston. Make sure that the trees you purchase are grown from seed that has been collected from trees with appropriate provenance (region of origin) for your climate. One way to insure this is to ask your nursery if the trees were grown from locally collected seed.

Cold hardiness is also an issue to be aware of when purchasing cultivars, because these selections were all made originally from trees of differing native provenance. The hardiness of the cultivar will be close to that of its parent population of trees. For example, the cultivar 'Royal White' was found in the wild in Illinois and so is very cold-hardy, while the cultivar 'Wither's Pink Charm' was found in the mountains of Virginia and will not be as cold-hardy as 'Royal White'.

Climatic factors as well as the provenance of the tree can influence the hardiness of individual redbud trees in your garden. When redbud is grown in environments with high light and relatively warm summers (e.g., most of the United States, except for the Pacific Northwest), it will survive lower temperatures and colder winters than when it is grown in climates with cool, cloudy summers (e.g., England, or the Pacific Northwest). This is why you will find English gardeners and texts complaining that eastern redbud will not survive their winters, although the winters in most of England are far milder than the winters that most eastern redbuds routinely survive each year in North America. The lack of light and the cool temperatures during the growing season impair redbud's ability to achieve the degree of cold hardiness that it can under brighter, warmer conditions.

Redbud is something of a challenge to propagate. It is more readily propagated from seed than vegetatively, but even so, seed must be pretreated with an acid soak and then stored in moist peat at 40°F (4°C) for 2 months before it will germinate. Cultivars must be grafted. Successful grafting of cultivars and individual selections is very difficult and a challenge for even professional grafters.

The following cultivars and varieties are all selections of eastern redbud (*Cercis canadensis*). For an excellent discussion of all of the species of *Cercis*, including the cultivars of the Asian species, refer to Dr. J. C. Raulston's extensive review article of the genus in *American Nurseryman* (171(5): 39–51). A white-flowered form, 'Alba', or var. *alba*, occurs relatively frequently in nature and will come true from seed. The clear white blooms are a refreshing change, and the variety is beautiful in combination with the more common pink-flowered form. The foliage of 'Alba' is usually a lighter green than the straight species. Cold hardiness can vary among populations of white-flowered eastern redbud, since the white-flowered form occurs spontaneously in many native populations.

A number of other good cultivars are available. 'Appalachian Red' has intensely colored flowers the closest to true red of any other named forms (Plate 8). 'Flame' is a semidouble to double-flowered form found in the wild in Illinois and should be quite cold-hardy. 'Forest Pansy' is one of the most popular cultivars, with deep burgundy foliage and rose pink flowers. New foliage emerges black-purple, turning burgundy red as leaves expand. The wine-purple foliage retains its color well in climates with cool summers but lightens to green in the hot summers of the Southeast and Southwest. It is likely only hardy through zone 6. 'Pinkbud' has bright pink flowers. The parent plant is from Kansas City, so this cultivar should be relatively cold-hardy. 'Royal White' is a white-flowered form with larger, more abundant flowers than 'Alba'. It was selected in Illinois and is one of the most cold-hardy cultivars of eastern redbud. 'Silver Cloud', a spectacular selection, has leaves that are variegated with large splashes of white to creamy white (Plate 9). The variegation is best in cool climates, and in warm climates it lasts well into the summer but eventually fades to light green. It produces very few flowers. 'Withers Pink Charm' has flowers which are a true soft, baby pink, without the rose-purple hues of the species. It was found in the mountains of Virginia and is likely only reliably hardy through zone 6.

Cultivars of other species of redbud are also exceptional. *Cercis canadensis* ssp. *texensis* is the species name for populations of redbud in Oklahoma, Texas, and Mexico, with distinctive, leathery, glossy, bright green leaves. *Cercis canadensis* ssp. *texensis* 'Oklahoma' is one of the most dramatic redbud cultivars available (Plate 10). It produces brilliant red-purple flowers in great abundance and has leathery, glossy, bright green leaves (Plate 11). The flower display on this cultivar is a showstopper, and the foliage remains exceptionally beautiful throughout the summer. It was found in the mountains of Oklahoma and should be hardy through at least zone 6. *Cercis canadensis* ssp. *texensis* 'Texas White' is a white-flowered form of the *C. canadensis* ssp. *texensis* type, with the leathery, glossy, bright green foliage of this race and abundant, bright white flowers. It was found in Texas and should be hardy at least through zone 7 (and possibly zone 6).

Cercis chinensis is one of the species found in China and can be larger and coarser in size and habit than North American redbuds, reaching 50 feet (15.2 m) in its native habitat. In cultivation in the United States, however, it is usually closer in stature to native redbuds. Growth habit of this species is significantly more upright than that of other redbuds. The cultivar *C. chinensis* 'Avondale' is a dark-flowered form with deep lavender flowers that are densely borne along all the branches, to the point where, often, almost no bark is visible. Hardiness should be reliable through at least zone 6.

No other small flowering tree has the distinctive charm and unforgettable color of redbud. Whether you choose one of the excellent cultivars, perhaps as a stunning specimen canopy for rare spring bulbs, or scatter seedlings along the edges of your streams and little woods, redbud will soften your spring garden with its lovely cloud of pink. It is a tree that no garden should have to do without.

Chaenomeles speciosa / Flowering quince   PLATE 12

A hedge of flowering quince in full bloom can be an enthusiastic burst of coral in the gray days of early spring. One of the first to greet spring, this unassuming shrub blooms delightfully any time from January through April, depending on the region. The gray-brown branches are studded with waxy flowers in opaque shades of orange-red, rose, pink, or white. Individual flowers are borne right against the branches and are reminiscent of apple blossoms, but their waxiness lends them a stouter character.

This deciduous, broadly spreading shrub grows from 6–10 feet (1.8–3.1 m) tall and can spread as wide. It is a twiggy and tangled multistemmed plant that makes a good barrier or hedge but which can also be a trap for dead leaves and other debris. Flowering quince will tolerate a wide range of soil and site conditions, including dry areas, but will suffer if planted in high pH soils. These shrubs are hardy to zone 4, but they do not perform well in the more "tropical" areas, such as southern Florida. Flowering quince is susceptible to fireblight disease and should therefore not be planted where it could infect fruit trees in a commercial or home orchard. Leaf spot can also be a problem during wet periods, but although it may cause some defoliation, it rarely leads to any permanent damage.

For best growth and flowering, plant flowering quince in full sun and prune out the old canes and suckers every year. If flowering is sparse, try pruning the entire plant down to 6–12 inches (15.2–30.4 cm) from the ground. Pruning should be done in March to May, after flowering, so as not to damage the flowering potential for the following year. Flowering quince can be propagated from softwood cuttings. Taken in summer, the cuttings can be treated with rooting promoters and rooted under mist. Flowering quince gradually spreads into clumps by suckering, and the suckers can be dug and divided like herbaceous perennials to obtain additional plants.

Spring flowers are the flowering quince's best feature. They have a unique, fresh appeal as they are among the earliest flowers to appear. New leaves emerge a bronzed reddish color that turns a glossy green, but fall color is unremarkable. There is no significant winter interest but the densely twiggy character of the plant makes a good shelter for birds and other small wildlife (while successfully keeping out the neighbor's dog). Flowering quince is a relative of the edible quince, *Cydonia oblonga*, and does produce small, quince-like fruits, albeit in a somewhat miserly fashion. The fruits are truly awful when eaten fresh but do make tasty preserves.

A number of flowering quince cultivars hold special interest. 'Cameo' (Plate 12) bears lovely double peach-colored blooms in great profusion. 'Spitfire' is an upright form with bold, bright red flowers. 'Contorta' has a contorted growth habit that adds greatly to the winter interest of this cultivar. Flowering quince is readily available in many nurseries and garden centers where it offers an inexpensive and undemanding treat of new spring color.

Chaenomeles is a bit of an old-fashioned workhorse in the garden. It is not rare, nor does it offer year-round excitement. Instead, it faithfully braves the unpredictable early spring to give us a welcome greeting of sprightly color.

Chionanthus spp. / Fringetrees   PLATE 13

The flowers of fringetrees create a marvelous haze of creamy white bloom suspended from the branches. The display is very dramatic, somewhat reminiscent of a white version of smoketree (*Cotinus*). Best known is the American fringetree, *Chionanthus virginicus*, which is indigenous to most of the southeastern United States. It can generally be found along the banks of streams and ponds or in other areas with moist, rich soil, although it has been reported in other types of soils as well. There is also a lesser known native fringetree, *Chionanthus pygmaeus*, dwarf fringetree, a more diminutive, slower growing form. It is found only on the sandy soils of Florida and is listed as an endangered species. Our native fringetrees are not the only type available to gardeners. Chinese fringetree, *Chionanthus retusus*, was introduced to this country in the 1800s. A native of China, Japan, Korea, and Taiwan, it is similar to American fringetree but has special qualities of its own, such as its leathery, rounded foliage. The fringetree genus name was coined by Linnaeus, the Swedish botanist responsible for our system of technical botanical names, who named them *chionanthus*, from the Greek for "snowflower" (*chion*, "snow," and *anthus*, "flower").

American and Chinese fringetrees make excellent landscape plants. They are small deciduous trees, or sometimes multistemmed large shrubs. American fringetree will generally reach 15–20 feet (4.5–6.2 m) in the landscape but can get to 35 feet (10.9 m) with age. It has a somewhat open and gangling branching habit with an overall broadly rounded to oval shape. In the spring, many-flowered panicles of small blooms with creamy white, straplike petals drape the branches, generally before the leaves emerge. Flowers can also emerge later, after plants have leafed out. Each flower alone is not especially showy, but together, they create a unique cloud of bloom. The fruits that follow can also be very beautiful. They are bluish and look like tiny plums. They shrivel somewhat late in the season. Fringetrees have male and female flowers on separate plants, like hollies, so a female plant, with males in the vicinity, is needed to have fruits. Usually, though, the male trees have more profuse flowering displays because there tend to be more flowers per plant.

The foliage of American fringetree is an attractive glossy green and begins to come out as the flowers mature to yellow-cream. Fall color is yellow but is not especially showy. American fringetree's bark is an attractive light gray and becomes ridged as the plant matures.

Chinese fringetree also blooms in the spring, a bit later than American fringetree. It tends to have smaller, more compact flower clusters and hold them a bit more upright on the branches. The bloom color is closer to a clear white and gives a more clustered, opaque effect to the flowering display, as if the tree were covered in pure white balls of fleece. The leaves of Chinese fringetree are smaller and more leathery and shiny than those of the American fringetree.

Chinese fringetree has the potential to reach 40 feet (12.4 m), but it is usually seen in the range of 15–25 feet (4.5–7.5 m) in height. It has a broadly spreading habit and more regular branching than its American cousin. In general, Chinese fringetree gives a

PLATE 2. *Aesculus* 'Briottii' in flower is one of the loveliest small trees for the garden.

PLATE 1. Not only are the tall spikes of *Aesculus parviflora* flowers appealing but the foliage and habit are equally as handsome.

PLATE 3. The fruits of *Akebia* split as they ripen to reveal the seed inside.

PLATE 4. The large, fleshy fruits of *Akebia quinata* make a delightfully unusual display hanging from a trellis.

PLATE 5. Spring-blooming *Amelanchier* are naturals to combine with spring-flowering bulbs.

PLATE 6. *Bignonia capreolata* 'Tangerine Beauty' is a crossvine of a different color with deep tangerine colored flowers.

PLATE 7. The ruffled flowers of catalpa trees are a wonderful late spring garden ornament.

PLATE 8. The early spring display of eastern redbud (*Cercis canadensis* 'Appalachian Red' pictured here) in full bloom is one of the loveliest of all small native trees, in the landscape or in the wild.

PLATE 9. The foliage of *Cercis canadensis* 'Silver Cloud' is heavily variegated with creamy white.

PLATE 10. *Cercis* 'Oklahoma' has some of the most deeply colored and prolific flower displays of all the redbud cultivars available.

PLATE 11. The foliage of *Cercis canadensis* ssp. *texensis* 'Oklahoma' is especially rich green and glossy with an undulate margin.

PLATE 12. 'Cameo' is a double-flowered pink selection of flowering quince (*Chaenomeles speciosa*).

PLATE 13. Chinese fringetree (*Chionanthus retusus*) has especially dense, compact inflorescences.

PLATE 14. The flowers of Armand clematis (*Clematis armandii*) are clear white and sweetly fragrant.

PLATE 15. The exfoliating bark of *Cornus mas* is highly ornamental.

PLATE 16. *Cornus mas* is covered with bright yellow blooms each spring.

PLATE 17. Chains of buttercup-yellow flowers make *Corylopsis pauciflora* a must for spring gardens.

PLATE 18. The bright flowers of *Cytisus* 'Lena' are dramatically showy.

PLATE 19. The creamy flowers of 'Moonlight' make this one of the most desirable selections of *Cytisus* available.

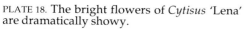

PLATE 20. The famous flowering display of dove tree (*Davidia involucrata*) is made up of showy white bracts draping small centers of inconspicuous true flowers.

PLATE 21. *Enkianthus campanulatus* bears delicate, bell-shaped flowers tinged with rosy-pink.

PLATE 22. *Enkianthus cernuus* var. *rubens*, with its deep rose-red flowers, is among the most highly colored of the genus.

PLATE 23. *Fothergilla gardenii* produces intriguing displays of bottlebrush-like clusters of flowers with no petals, but long showy stamens.

PLATE 24. The luscious flowers of *Gardenia jasminoides* have an unforgettable, exotic fragrance.

PLATE 25. *Gardenia jasminoides* 'Variegata' has foliage generously splashed with creamy white to gold.

PLATE 26. The bright yellow flowers of *Hamamelis* 'Arnold Promise', a selection from the Arnold Arboretum, are delightful in early spring.

PLATE 27. *Hamamelis* 'Ruby Glow' is a striking hybrid selection with red flowers.

PLATE 28. 'Nellie R. Stevens' holly is reliable in hot areas and produces some bright red fruit even if a male pollinator is not nearby.

PLATE 29. The starlike flowers and oval, ever-green foliage of *Illicium anisatum* make this an excellent ornamental shrub.

PLATE 30. Tough and attractive groundcover juniper (*Juniperus horizontalis*) lends itself to many creative uses in the garden, including as a trailing evergreen planted on top of a wall.

PLATE 31. 'Blue Star' is a cultivar of single-seed juniper (*Juniperus squamata*) selected for its bright blue foliage and compact habit.

PLATE 32. Chinese loropetalum is covered with snowy, fine-textured flowers in spring.

PLATE 33. There are several new, exciting cultivars of Chinese loropetalum with amazing magenta flowers.

PLATE 34. The chalice-shaped flowers of Yulan magnolia are among the most beautiful blooms of all the deciduous magnolias, and their fragrance is exquisite.

PLATE 35. Star magnolias (*Magnolia stellata*) are prolific bloomers in diverse climates and growing conditions.

PLATE 36. *Magnolia stellata* 'Rubra' has petals with deep rose colored outer surfaces and pale pink insides.

PLATE 37. The flowers of *Magnolia virginiana* give a more refined display than those of its larger, coarser cousin, *Magnolia grandiflora*.

PLATE 38. The showy salmon-pink bracts of *Pinckneya pubens* create a display like a tree full of pink poinsettias.

PLATE 39. *Prunus glandulosa* 'Alba Plena', a double, white-flowered selection of dwarf flowering almond, is an economical source of foolproof display each spring.

PLATE 40. 'Okame' cherry (*Prunus × incamp*) is one of the earliest blooming, and easiest to grow flowering cherries and useful with many different garden styles.

PLATE 41. 'White Glory' weeping nectarine (*Prunus persica* var. *nucipersica*) is a waterfall of white bloom each spring.

PLATE 42. The flowers of 'Yoshino' cherry (*Prunus × yedoensis*) dangle in airy, cloudlike clusters with a delicate, shell-pink cast.

PLATE 43. The large, soft catkins of *Salix gracilistyla* make it one of the showiest of the pussy willows for spring gardens.

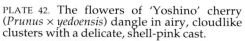

PLATE 44. *Styrax japonica* 'Carillon', a heavily flowering selection of Japanese snowbell, also offers especially dark green foliage and near-black pedicels.

PLATE 45. Korean early lilac (*Syringa oblata* var. *dilatata*) combines the appeal of common lilac with exceptional heat tolerance and reliable performance in heavy soils.

PLATE 46. 'Eskimo' is one of the most ornamental of a series of *Viburnum carlesii* hybrids with large, snowball inflorescences. 'Eskimo' was recently chosen as one of the Pennsylvania Horticultural Society Gold Medal award winning plants.

PLATE 47. *Viburnum tinus* 'Compactum' is a more compact growing form of the species with an equally attractive display of pink flower buds and pinkish white flowers.

PLATE 48. Weigela 'Rubigold' is a brazen culti-var with bright lime-gold leaves and magenta flowers.

PLATE 49. Few vines can top the dramatic impact of wisteria in bloom on a trellis.

PLATE 50. The velvety fruit of *Wisteria* extends the ornamental season of this vine.

PLATE 51. Dusty zenobia (*Zenobia pulverulenta*), a native shrub, has an exciting flowering dis-play and exceptional fall color.

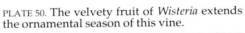

neater, more orderly, and more compact appearance, both in flower and throughout the season. The bark of Chinese fringetree begins to peel in an interesting manner as the plant matures. Like the American species, Chinese fringetree bears male and female flowers on separate plants and produces showy dark blue fruit on the females.

The American and Chinese fringetrees are easy, stress-tolerant trees for sunny sites. They have no disease or pest problems and will thrive in most landscape settings. American fringetree is the more cold-hardy of the two (to zone 3), but Chinese fringetree is quite cold-hardy as well (to zone 5). They are both grown from seed. Seed from both have "double dormancy," which means that the seed requires several different treatments before it will germinate (in nature, the seed has two "rest periods," usually requiring two years to sprout). Because the plants are grown from seed, the performance and appearance of individual trees varies greatly.

There are no cultivars of fringetrees because vegetative propagation is practically impossible. Chinese fringetree can be propagated with difficulty from rooted cuttings, but American fringetree cannot be propagated from cuttings at all. They are sometimes grafted onto ash (*Fraxinus*) rootstock, but the resulting plants are not very vigorous. Nonetheless, both species of fringetrees are available from specialty nurseries. The American fringetree in particular is becoming easier to find, as more nurseries expand their offerings of native plant materials.

Fringetrees offer unique drama in the garden. Plant hunter and explorer Frank Meyer, in 1915, described a Chinese fringetree in full bloom as looking "as if white muslin cloth had been thrown over its head"—an apt description of an exceptionally floriferous plant for modern gardens. In the Southall garden at the North Carolina State University Arboretum, Chinese and American fringe trees stand side by side and offer a rare opportunity for fringetree lovers to closely compare their spectacular shows.

Clematis armandii / Armand clematis   PLATE 14

With their ability to climb vertical surfaces and cover horizontal spans, vines are capable of transforming the commonplace landscape into a spectacular, multidimensional garden. Chief among the many gardenworthy vines are the hundreds of species and hybrids of showy-flowered clematis. But even among so many clematis choices, *Clematis armandii*, the Armand clematis, stands out as an exceptional flowering vine.

Armand clematis is an evergreen climbing vine native to China. Its dark, leathery leaves are very large. Individual leaves are oval or somewhat heart-shaped and can reach 6 inches (15.2 cm) long and 2 inches (5 cm) wide. The foliage drapes from the woody, twining stems all year and keeps good color throughout the winter. In early to mid-spring, the vine is covered with charming white, almost translucent, flowers that age to a very light pinkish color. Unlike the blooms of the more common large-flowered clematis hybrids, individual flowers of Armand clematis are not exceptionally large, usually about 2 inches (5 cm) across. Although individual flower size is demure, many flowers are in bloom at the same time, and they create a lacy floral quilt that covers the vine with delicate beauty in early spring.

The individual flowers are also lovely by themselves. The showiest part of clematis flowers are the sepals. The fruit of clematis are also interesting. Brown, very small seedpods with long, soft hairs develop through the summer. While these silky pods are not especially eye-catching from a distance, they are an attractive reward for the attentive gardener in the fall.

Culture of Armand clematis is not difficult but does require more awareness than some plants. Clematis in general prefer cool, moist, well-drained, loamy soils where the foliage can climb up into a warm, sunny perch. These conditions will result in the best flowering. Less than perfect sites can also give very good results, if the roots are kept cool by planting in a partially shaded area, and perhaps using a light mulch. Partial shade is also helpful in preventing the evergreen leaves from becoming sun-scorched in winter.

Clematis can be propagated from seed, which requires a long chilling pretreatment, or from softwood cuttings taken in the summer and rooted under mist. Armand clematis is reliably hardy through zone 7. Temperatures below 0°F (–16°C) may completely kill the plant.

Armand clematis is a rather uncommon plant but it is well worth a bit of hunting. There are some cultivars, including 'Farquhariana', a selection with rare true-pink flowers. 'Apple Blossom' has pink-tinted flowers and bronzy new foliage. 'Snowdrift' has especially clear white blooms. 'Early Spring' is an early bloomer with pale pink flowers.

Armand clematis will cover a trellis, a fence post, or a bannister with a blanket of delicate floral beauty in the spring and handsome, dark green leaves throughout the year. Look for this lovely vine in specialty nurseries so that you can bring its appealing charm into your own garden.

Cornus mas / Cornelian-cherry dogwood   PLATES 15, 16

The name "dogwood" conjures up images of clouds of white flowers drifting against the pinks and reds of azaleas and redbuds. Those classic white blooms belong to the native flowering dogwood, *Cornus florida*, only one of many different and excellent dogwoods, many not as well known as the favorite native, but equally deserving of attention. One such plant is the cornelian-cherry dogwood, *Cornus mas*, a native of western Asia that has been cultivated in Asia and Europe since ancient times.

Cornelian-cherry dogwood has two special characteristics that set it apart: its flowers and its fruits. It flowers in early spring before most other flowering trees and shrubs are out, bearing bright yellow flower clusters that almost glow against the rich brown bark. By July, those flowers have matured to fire-engine-red fruits. Although the edible fruits taste a bit like sour cardboard to many people, they are favorites of birds and other wildlife.

Cornelian-cherry dogwood is a multistemmed small tree with ornamental flaking bark, which will grow to 20 feet (6.2 m) in height and spread to 15 feet (4.5 m). The foliage is much like that of our old friend, the flowering dogwood, but it is more refined and, unlike flowering dogwood, offers little in the way of autumn color. The plant can be treated as a large shrub or a small tree in the landscape, especially if pruned to remove the lower branches.

Cornelian-cherry dogwood is hardy through much of North America, to zone 4. It transplants easily and is fairly adaptable to different soils, although it does best in rich soils and full or partial sun. This tree has no serious pest or disease problems. It can be propagated by seed or vegetatively. To propagate by seed, the seeds are placed in slightly damp peat moss in a plastic bag and stored at room temperature for four months. Then the bag is placed in a refrigerator for an additional four months before sowing. For vegetative propagation of selected forms, softwood cuttings can be taken in June and July.

A number of cultivars of this tree offer additional character in the landscape. 'Alba' is a white-fruited form, while 'Flava' bears golden fruits. 'Aurea' has golden foliage, and 'Variegata' is an unusual variegated cultivar with cream-colored leaf margins. 'Golden Glory' flowers very profusely and has a somewhat more upright and treelike form than other cultivars. 'Nana', on the other hand, remains low and compact and has small leaves. The North Carolina State University Arboretum, with the North Carolina Association of Nurserymen, is releasing an especially showy cultivar, 'Spring Glow', which is covered with a solid mass of gold flowers extraordinarily early each spring.

Cornus mas is one of the few early-flowering shrubs or trees that are hardy enough to brave the first frosts of northern springs. This sprightly little tree has much to offer southern landscapes and gardens as well. Cornelian-cherry dogwood is tough, fairly adaptable, and reliably beautiful in flower well before the more familiar flowering dogwood has begun to show even a hint of white or pink. The curious, attractive fruits bring a bright red note to the green foliage of midsummer. It is unlikely that you'll give up sweet cherries for cornelian cherries, but the fun of trying a taste is yet another reason to plant *Cornus mas* in your garden.

Corylopsis spp. / Winter hazel   PLATE 17

Each year, when spring begins to claim the garden, winter hazels (the many species of *Corylopsis*) unfurl pale gold chains of buttery, rounded, bell-shaped flowers. These pendulous flower clusters are made up of numbers of individual blooms draped in groups, as if unopened buttercup flowers had been strung together in braids 2–6 inches long (5–15.2 cm), and hung from the branches. The visual effect is completely delightful, but this is not the only attraction of winter hazel. The flowers are also fragrant, with a spicy-sweet scent reminiscent of witch hazel (*Hamamelis*), a close relative.

The genus *Corylopsis* includes a number of species, all of them deciduous multi-stemmed shrubs that bear clusters of pale gold to deep yellow flowers in spring, before the leaves emerge. They vary in height, showiness of flower clusters, and foliar characteristics. The medium-green foliage is coarse-textured and generally covered with a light down on the undersurface. Fall foliage varies from a lovely clear yellow to nondescript. The leaves sometimes remain green on the branches until killed by the first hard freeze. The light tan bark is beautiful with the sunny-colored flowers.

Cold-hardiness varies among the different species of *Corylopsis*, with *C. glabrescens* being the most hardy (zone 5) and the most reliable in northern areas. The other species listed here are reliably hardy through zone 6. Because they are early bloomers, flowers and leaf buds of *Corylopsis* may be damaged by late freezes, especially in areas where prolonged spells of warm winter weather, which may speed flowering, are followed by severely cold weather (e.g., northern areas of the southeastern United States). *Corylopsis* prefers well-drained, moist, acidic soils, but a number of different species also thrive in clay.

Corylopsis flowers most profusely in full sun, but the shrubs will also perform well in light shade, where they may benefit from the slight frost protection of a light overhead canopy. *Corylopsis* is easily propagated vegetatively from cuttings taken in summer and rooted under mist after treatment with rooting promoters. Newly rooted cuttings have sensitive root systems. To insure survival, they should be left undisturbed in their rooting medium to rest through one winter, and then they can be successfully potted up the following spring. Propagation of *Corylopsis* from seed is difficult, requiring long-term temperature treatments over several months.

Corylopsis glabrescens, fragrant winter hazel, is probably the most readily available winter hazel. This robust native of Japan is the hardiest of the group and will reach 8–15 feet (2.4–4.5 m) in height with similar spread. This species can be limbed up to create a small, rounded, multistemmed tree. Flowers are pale yellow with a light, sweet scent and are borne in 2-inch-long (5 cm) clusters before the emergence of the very large leaves, which are 4 inches long (10.1 cm) and 3 inches (7.6 cm) wide. This species is sometimes listed as *C. gotoana*; taxonomists are not in complete agreement as to whether these are actually two separate species, with *C. gotoana* being exceptionally rare, or are two names for the same plant. But by any name, fragrant winter hazel is a delightful addition to the garden.

Corylopsis pauciflora, buttercup winter hazel, is also native to Japan, as well as Taiwan, but is a much more demure creature, only reaching about 5 feet (1.5 m) in height with a spread of 5–6 feet (1.5–1.8 m). The leaves are a bit more refined in shape and texture and are smaller, generally 2 inches (5 cm) long and 1–2 inches (2.5–5 cm) wide. The bright yellow flower clusters are short, 1–2 inches (2.5–5 cm). This species is much more sensitive to high pH soils and winter full-sun exposure than its more vigorous cousin. It performs best in light shade and is perhaps the best choice for small suburban settings with existing large trees. Like all *Corylopsis*, it is very beautiful when in bloom against a dark background of evergreens.

Corylopsis platypetala has no acknowledged common name. It can grow to be quite large, reaching 15 feet (4.5 m) or more in a favorable site. Pale, creamy yellow flowers bloom on 2-inch-long (5-cm) clusters and are very beautiful. This native of China is tolerant of full sun or light shade but needs adequate space.

Corylopsis spicata, spike winter hazel, is a native of Japan with a very open, bushy habit enhanced by spreading, picturesque branching. It is frequently wider than it is tall, reaching 6–10 feet (1.8–3.1 m) in height with at least that much spread. Flowers are clear yellow and borne on clusters 1–2 inches (2.5–5 cm) long. The foliage of this species is exceptional, as the new leaves open a dark purple and mature to a deep blue-green.

Corylopsis willmottiae, Willmott winter hazel, is a native of China rarely found in the United States. This species probably has the largest flower clusters of the *Corylopsis*. The clusters may be 3–5 inches (7.6–12.7 cm) long and are covered with pale, green-yellow blooms. The shrub itself can reach 12 feet (3.7 m) in height, with a spread of up to 10 feet (3.1 m) or so. The cultivar 'Spring Purple' has reddish purple young shoots that change to green as the season progresses.

Like deciduous magnolias, winter hazels brave the first warm days of spring with a magnificent display that may suffer from late freezes. But the sight of their buttery flower clusters hanging in perfect golden chains (most years) is worth the disappointment of an odd year with late harsh weather. Including these delightful shrubs in your landscape will add a new horticultural treat as you walk your garden to find clusters of glowing blossoms waiting around each bend.

Cytisus scoparius / Scotch broom   PLATES 18, 19

The name "Scotch broom" conjures up images of broad, lonely moors, with wind crying eerily around a large moldering mansion occupied by dark, brooding people embroiled in lives more complicated than any modern soap opera ever dreamed of—the stuff of Victorian novels. But in fact, *Cytisus scoparius*, Scotch broom, is very real and delightfully cheerful. This tough shrub is perfect for modern gardens on both sides of the Atlantic.

Scotch broom is a multistemmed, fine-textured shrub with many slender, upright arching stems that are bright emerald green 12 months of the year (similar to those of winter jasmine, *Jasminum nudiflorum*). The bright green leaves are divided into three lobes, which gives the stems the appearance of being covered with hundreds of narrow clover leaves. *Cytisus scoparius* will reach anywhere from 5–10 feet (1.5–3.1 m) in height with a spread of 5–8 feet (1.5–2.4 m), depending on the site. The bright green stems add life to the winter garden and are a graceful addition to a mixed border during the summer, especially as they retain their good, fresh green color through the tail end of summer when many other shrubs have begun to look tired.

Scotch broom shines in late spring when the stems are completely covered with brilliant gold, pink, or red blooms (depending on the cultivar). Similar in shape to sweet-peas, the flowers can be so dense that they appear to be spontaneous rockets of bloom shooting from the ground. There is absolutely nothing "brooding" about Scotch broom in flower.

Cytisus scoparius is a tough plant in the landscape. Native to the dry soils of central Europe, it will do very well in poor, sandy soils where other showy shrubs remain stunted and suffering. In addition, this shrub also performs well in heavier clay soils. It is a good candidate for many tough sites, especially where salt and drought tolerance are required. In fact, it is so well adapted to dry, poor soils that regular water and fertilizer can actually shorten its life. Scotch broom needs full sun. It is intolerant of root disturbance and should be transplanted as a container-grown plant. If pruning is desired, stems should be cut shortly after flowering is finished. Scotch broom is completely cold-hardy through zone 5. It can be propagated from stem cuttings taken during summer, treated with high concentrations of rooting promoter, and rooted under mist.

Scotch broom is a brilliant burst of color in the landscape during its peak display in the spring, but it also blooms sporadically throughout the summer and into fall, adding touches of color for months. It is an absolute showstopper in combination with other flowering shrubs and perennials, and its form and color are marvelous against stone walls.

Cytisus is a very popular garden plant in Europe, where scads of cultivars are available. In North America, the plant is much less common and only a few of the cultivars turn up on a regular basis. More selections are being produced, and it is well worth the trouble to look for the wider range of colors available in the many named forms.

Cultivars include 'Burkwoodii', with light red flowers; 'Carla', light pink flowers lined with white; 'C. E. Pearson', rose, yellow, and red flowers; 'Cornish Cream', cream to buff blooms; 'Dorothy Walpole', rose flowers with deep red "wings"; 'Firefly', two-tone flowers of gold and deep burgundy; 'Goldfinch', yellow-and-scarlet flowers; 'Johnson's Crimson', scarlet blooms; 'Killiney Red', a dwarf form with bright red flowers; 'Knaphill Lemon', lemony yellow flowers; 'Lena' (Plate 18), multicolor ruby-red and light yellow flowers; 'Minstead', magenta-and-white flowers; 'Moonlight' (Plate 19), creamy yellow to white blooms; 'Nova Scotia', a very hardy form with brilliant yellow flowers; 'Red Favorite', a compact form with fire-engine-red flowers; and 'Zeelandia', with pink flowers on especially long, arching branches.

A great many other species and hybrids of *Cytisus* may be found with some hunting. You might encounter *C. procumbens* and *C. decumbens*, both of which are aptly named "groundcover types," with bright gold flowers. The dwarf *C. hirsutus*, hairy broom, is the most cold-hardy of *Cytisus*. *Cytisus multiflorus* is a white-flowered broom, while *C.* × *praecox*, Warminster broom, is a hybrid of complex parentage with a number of named forms in a range of colors.

Two closely related genera are *Genista* (woadwaxen) and *Ulex* (gorse). *Genista*, like

Cytisus, includes a number of species and cultivars. *Genista* is very similar to *Cytisus*, but finer and spikier in texture. *Ulex*, gorse, is a somewhat coarse and unbelievably spiny shrub, similar to *Cytisus* in most other aspects, except that it usually flowers earlier than *Cytisus*. *Ulex europaeus* is generally the only species of gorse seen in gardens.

Cytisus can be dazzling in many types of gardens. Broom is easy to see, even from a distance, and the masses of bloom stand up well to intense summer sunlight. They are a bright glow of color in the landscape—a familiar American landscape, without even a hint of moor in the distance.

Davidia involucrata / Dove tree   PLATE 20

Dove tree, *Davidia involucrata*, is one of the most famous, beloved, revered, and agonized-over trees among serious horticulturists in the temperate areas of the world. Famed as the first of many quests of renowned plant collector Ernest "Chinese" Wilson, who collected seeds of this magnificent tree in China in the 1800s, *Davidia involucrata* is revered for its otherworldly display of large white flower bracts, which flutter like wings of doves from the branches in spring. It is agonized over for the decade or so that it takes a young tree to reach flowering age, but no gardener ever regrets that wait when the first flowers finally open.

Davidia involucrata is named for a Frenchman, Armand David, who first found and brought seed back from a tree in China in the early 1800s. The seed from David's collection, however, resulted in only one surviving tree, and so the European plant community could not distribute any other plants. Later in the 1800s, Ernest Wilson, the famous collector affiliated with the Arnold Arboretum, went to China on an expedition devoted to finding the dove tree. He collected a great deal of seed, which was successfully grown into hundreds of seedling trees, many of which were distributed among the horticultural community at that time. This is why any discussion of the dove tree is more likely to include reference to Wilson rather than the tree's actual western discoverer and honoree by name, Armand David.

The reason these Victorian gentlemen were so obsessed with *Davidia* is the same reason it drives modern horticulturists to flights of ecstasy (and depths of despair). The tree's small clusters of tiny flowers are wrapped in two large, delicate bracts. One bract, 2–3 inches (5–7.6 cm) long, drapes from the top of the cluster, and a second, longer bract, 4–7 inches (10.1–17.7 cm) long, drapes straight down from the bottom. The two bracts together resemble the wings of a dove (albeit somewhat uneven wings), or a folded handkerchief fluttering in the breeze (hence *Davidia*'s other common name of 'handkerchief tree'). A tree with prolific bloom will be covered in these dramatic bracts. The flowers hang from long pedicels and the bracts are easily blown by any light breeze—such a tree looks as if it could sail up off the ground at any moment.

Davidia involucrata is an attractive large tree, reaching 50–60 feet (15.2–18.6 m) in height, generally with a rounded pyramidal habit and dense branching. The large leaves are an appealing medium green. The foliage does not develop any appreciable fall color.

Seed-grown trees vary in their habit and flower characteristics. Final mature habit varies considerably among seedlings, from somewhat squat or broadly rounded to quite upright and conical. Seedlings differ in the size of flower bracts and degree of bloom. The most spectacular individuals produce flowers that cover the tree with 8–10-inch (20.3–25-cm) bracts, while lesser individuals may show few blooms with bracts only 4–5 inches (10.1–12.7 cm) long.

Even the most modest seedlings, however, convey the magic of these trees to the garden, so why isn't dove tree in every landscape? The problem is that trees often take ten years to bloom, and then they may develop an alternate-year bloom pattern (as many crabapples do), so that they are only showy every other spring. The display of _Davidia_ is worth waiting for.

Dove tree is hardy to zone 6, but late frosts may damage flower bracts. It has no serious disease or insect problems and it does best in moist, well-drained, slightly acid soils in full sun or light shade. Regular moisture is important for good growth, and good growth is important in order for the tree to reach maturity and produce flowers. Dove tree is generally propagated from seed, which has a double dormancy requiring both warm and cold pretreatments. It can also be rooted under mist from cuttings taken in the summer. An especially large-bracted clone from Sonoma, California, has been successfully grafted as well. The botanical variety _Davidia involucrata_ var. _vilmoriniana_ is essentially identical to the species but has smooth undersides on the leaves and somewhat increased cold hardiness.

Davidia involucrata is one of the great horticultural treasures to come out of the heyday of plant collection expeditions to China. The vagaries and idiosyncracies of bloom only add to its irresistible appeal. Even the unflappable plantsman Michael Dirr reports, ". . . when observed in flower there is an insatiable urge to secure a plant for one's garden. . . ."

Dove tree is well adapted to modern landscape conditions, with the exception of very stressful urban sites, where it can become an unforgettable flight of white arboreal doves in the spring. No gardener's life is ever quite the same after finding a dove tree in full bloom. By the way, after 12 long years of waiting, the _Davidia_ at the North Carolina State University Arboretum finally bloomed in the spring of 1993.

Enkianthus / Enkianthus   PLATES 21, 22

Gardeners with an eye for the unusual can enjoy the refined beauty of _Enkianthus_, a group of shrubs that bloom after the flamboyant burst of dogwoods, azaleas, magnolias, and redbuds. All the enkianthus are multistemmed shrubs or small trees with delicate, bell-shaped flowers appearing in spring. There are a number of interesting species for the garden.

Enkianthus campanulatus, redvein enkianthus, is an uncommon but lovely plant with elegant, subdued character and four-season interest for the garden. Redvein enkianthus is a deciduous, upright-branching shrub or small tree, remaining at 5–8 feet (1.5–2.4 m) in height in colder areas but reaching to 12–15 feet (3.7–4.5 m) in the southeastern United States and other mild regions. The bark is a warm brown. Foliage is a glossy blue-green to olive color and remains neat from spring to fall. In autumn, the leaves can turn a vibrant red-orange or clear yellow. This trait is somewhat variable, so shop for plants in the fall and choose those that show good foliar color. In spring, this relative of the blueberry is hung with clusters of small, bell-shaped flowers as the foliage is emerging. The individual creamy-white bells are not very showy because they are only about 0.3 inch (12.2 mm) long, but the flowers are borne in clusters scattered throughout the plant. Upon close inspection, the flowers reveal red or pink veining near the edges of each blossom.

Enkianthus campanulatus is native to Japan, but it is well adapted to many sites in North America. It is a member of the same botanical family as rhododendron and azalea and, like those plants, does best in moist, well-drained, acidic soils. However, it is tolerant of other landscape conditions and has performed well in soils with significant amounts of clay and only moderately good drainage. Redvein enkianthus will do well in full sun or partial shade and has no serious pest or disease problems. The plant is cold-hardy to zone 4.

Redvein enkianthus can be readily propagated from seed, which has no pregermination requirements and can be sown directly onto moist peat moss. Softwood cuttings taken in summer root readily under mist, but the cuttings may not survive through the following winter. Commercial propagation of cultivars is frequently done using modern tissue culture techniques to avoid this problem.

A number of cultivars have been selected for exceptional flower and foliage color. 'Albiflorus' has clear white flowers with no colored venation, and orange-red fall color. The botanical variety *Enkianthus campanulatus* var. *palibini* has vivid red flowers. The blossoms of 'Red Bells' are deep red near the tips, and the venation of the flowers is a darker red than usual. 'Renoir' is a yellow-flowered form with pink tips, from the Arnold Arboretum. 'Showy Lantern', from Weston Nurseries in Massachusetts, has large pink flowers with excellent carmine fall color and exceptionally dense branching. 'Sikokianus' (sometimes listed as the botanical variety *E. campanulatus* var. *sikokianus*) has probably the darkest flowers with deep wine-colored buds that open into dark bronze-red flowers with lighter pink venation.

In addition to the cultivars of redvein enkianthus, several other lesser known species of *Enkianthus* offer unique character for the landscape. *Enkianthus cernuus* is a somewhat lower-growing shrub with pure white, bell-shaped flowers borne in clusters of 10 to 12 individual blooms. The botanical variety *E. cernuus* var. *rubens* has flowers of a unique deep red. *Enkianthus deflexus* is native to China and attains heights of up to 20 feet (6.2 m). Flowers are larger and showier than other species, but it is not quite as hardy as others and may suffer damage in areas colder than zone 6. *Enkianthus perulatus*, white enkianthus, blooms before the leaves are out, with creamy flowers, each 0.5 inch (1.2 cm) wide, that remain attached at the petal tips like small balloons. It remains lower growing than other *Enkianthus*, reaching only 4–6 feet (1.2–1.8 m) in height with a rounded habit. *Enkianthus serrulatus* is an uncommon species with the largest flowers of the genus. The waxy, translucent ivory blooms are 0.5–0.75 inch (1.2–1.9 cm) across. The plant is cold-hardy to zone 6. While *E. serrulatus* is perhaps the most beautiful of all the *Enkianthus*, it is also the earliest to bloom, making it most susceptible to spring frost damage.

At the North Carolina State University Arboretum, many of the different forms of *Enkianthus* can be seen in the shade house, the Southall garden and the white garden. Any of the *Enkianthus* are excellent plants used in small groupings around a patio or as specimens near a walk or seating area. Wherever they are grown, *Enkianthus* bring a refined sense of the unusual to the garden experience.

Fothergilla spp. / Fothergilla   PLATE 23

Azaleas, azaleas everywhere and no relief in sight! Azaleas can be very showy, massed in foundation plantings or used in combination with rhododendrons, but many interesting alternatives to our azalea addiction are easy to find if we only take the time to look. Fothergillas, *Fothergilla* spp., are another choice, very different from azaleas, that bring an unusual and sophisticated beauty to the garden.

Like azaleas, fothergillas make beautiful companion plants with rhododendrons. Unlike most azaleas, their white, feathery flowers do not blaze with intense color, and therefore they are easier to place in the landscape.

Fothergillas are deciduous small- to medium-sized native shrubs, related to witch hazels (*Hamamelis*) by virtue of belonging to the same botanical family. They are indigenous to the Southeast but are named after an Englishman, Dr. John Fothergill, who cultivated one of the earliest and most extensive collections of American plants in Europe.

Fothergilla's white flowers are borne in fine-textured "bottlebrush" clusters perched on the branches. The clusters are very feathery in appearance and are actually clusters of tiny flowers with no petals surrounding the interior parts of the flowers. The clusters are very feathery in appearance due to the lack of petals, which exposes the many fine white stamens. The peak display is usually in mid- to late April (or May, depending on the region) and lasts for one to two weeks. Fothergilla's irregularly rounded, toothed-margin leaves resemble those of its witch hazel relatives. The foliage has magnificent fall color ranging from gold to orange to scarlet, often on the same plant. Although *Fothergilla* can be grown in a range of conditions from shade to full sun, good fall color will develop only in full sun.

Two species of *Fothergilla* occur naturally in two different habitats and are also grown for the nursery trade. *Fothergilla gardenii*, dwarf fothergilla, is indigenous to the Coastal Plain of the Southeast. It is a slow-growing, rounded shrub generally reaching 3–5 feet (0.9–1.5 m) in the landscape (although it can very rarely attain 10 feet (3.1 m) with age). The flower clusters are 1–2 inches (2.5–5 cm) long and fragrant. The foliage is a lovely blue-green with a satisfying, leathery texture. A popular cultivar of dwarf fothergilla, 'Blue Mist', has been selected for its exceptionally beautiful blue-green foliage.

Fothergilla major, large fothergilla, is indigenous to mountainous areas and is taller and somewhat more upright than *F. gardenii*. Large fothergilla will generally reach 6–10 feet (1.8–3.1 m) in height and spread to nearly that width with age. The flower clusters tend to be somewhat longer than those of dwarf fothergilla, usually 2 inches (5 cm) long, and are also fragrant.

Fothergillas are essentially trouble-free plants that have no pest or disease problems. They are hardy to zone 4 and perform well in the same conditions that azaleas and rhododendrons require to do their best: moist, well-drained, somewhat acid soils; and partial shade in the Southeast and other areas with hot, humid summers and intense sun. *Fothergilla* is a good choice for a mixed planting with rhododendron. Propagation of the shrubs is difficult from seed as the seed has a double dormancy requirement. Vegetative propagation can be successful from softwood cuttings rooted under mist.

Fothergilla's intriguing brushes of flowers, attractive summer foliage, and mosaic of fall color are good alternatives to standard landscape plantings. A small mass of *Fothergilla major* in flower against rich green rhododendron foliage, or a specimen of dwarf fothergilla in a mixed border, perhaps as comrade to a special *Pieris*, are just two examples of the way *Fothergilla* can bring finesse to a garden.

Gardenia jasminoides / Gardenia   PLATES 24, 25

The fragrant elegance of a new gardenia blossom, open to the soft air of late spring and early summer evenings, brings the ambience of the Old South to the new walks and lawns of today's gardens. In hot, humid summers, they bloom in profusion, giving a marvelous gift to southern landscapes. But gardenias need not be restricted to southern gardens. They make lush container plants in a courtyard (to be overwintered indoors or under glass), or they can be used, like many tender perennials, as temporary transplants in the garden for the length of the season.

Gardenia jasminoides, the cape jasmine or common gardenia, was named to commemorate Alexander Garden, a doctor of some eminence in Charleston, South Carolina, who corresponded with the renowned Swedish botanist, Linnaeus. Native to southern China, gardenia is a rounded, broad-leaved evergreen shrub that reaches 4–6 feet (1.2–1.8 m) in height with approximately equal spread. Beautiful creamy white flowers with the exotic character of magnolia blossoms unfold from May through early summer. The 2–3-inch (5–7.6-cm) flowers are deliciously fragrant, with a heady perfume that is carried for a long time on the humid air of summer. The foliage of gardenia is as lovely as its flowers. Long, leathery, glossy leaves remain a deep emerald green throughout the year. The flowers may set orange, berrylike fruits late in the summer.

Gardenias make excellent specimen plants near walks, steps, gateways, or anywhere in the garden their fragrance can be appreciated. Porches and patios are also perfect sites for gardenias, not only because they are often the most sheltered areas of a landscape, but also because of the delight that comes from breathing the perfume of gardenias while rocking on a porch swing, counting fireflies. One of the most inspired uses of gardenias I know of was around a naturalized swimming pool, where a small collection of flowering gardenias was planted in a bed alongside a stone wall surrounding several teak chaise lounges. The heavenly fragrance wafted over guests as they reclined after a swim.

Gardenia jasminoides is hardy only to zone 8. In climates with colder winters, they can be enjoyed as container plants, which can then be moved and held in sunporches or other protected areas during the winter. Gardenias thrive in containers, and such potted plants give the gardener the advantage of being able to shift the pots to places where the blossoms and fragrance can be most readily enjoyed. In containers, make sure to use a light, well-drained medium but remember to keep plants evenly watered. The medium should feel moist, but not be puddled, at all times.

Gardenia thrives in well-drained, acid soil that is high in organic matter. They do well in full sun or partial shade, but should be protected from severe, desiccating winter winds. A few diseases and pests must be contended with, including white flies, canker, and powdery mildew. But these beautiful shrubs are worth the little bit of extra effort it takes to help them prosper. Site gardenia where air circulation is good, and remember that there pests and diseases will rarely have a significant impact on the plants' survival over time. Gardenia is easily propagated from softwood cuttings taken in summer and rooted under mist.

A number of cultivars of gardenia offer special beauty and drama. 'August Beauty' blooms very profusely with double flowers from May through September. 'Mystery', in addition to its magical name, has extra-large 4–5-inch (10.1–12.7 cm) double flowers. It can be more upright in habit than the species. The foliage of 'Variegata' (Plate 25) has cream-colored markings that add sophistication. 'Radicans' has a unique, almost prostrate habit, with small leaves and blooms of only 1 inch (2.5 cm) in diameter. It reaches only 2 feet (0.6 m) in height but spreads two to three times as wide to make a lovely dense mass. Both 'Variegata' and 'Radicans' are less hardy than the others and must be well protected from very low temperatures and extremes of winter weather.

In the early heat of late spring and early summer, the unfolding flowers of gardenia bring a touch of the languid South to gardens in all climates. The drifting fragrance adds another element to the pleasures of the season, inviting visitors to linger on the porch or follow the garden path to find the source of the enticing scent.

Hamamelis spp. / Vernal witch hazel   PLATES 26, 27

As you walk through the white garden at the North Carolina State University Arboretum, toward the columns that frame the path out into the rest of the garden, a delicious fragrance gently calls your attention. The planting that greets you as you pass the columns includes nandinas with their striking red fruit, some softly blue cypress, wintersweet, and the witch hazel collection. It is the winter-blooming witch hazels whose fragrance has enticed you past those columns.

Witch hazels are well known for blooming at odd times of the year—late autumn, winter, and earliest spring. Aside from common witch hazel (*Hamamelis virginiana*) and southern witch hazel (*H. macrophylla*), which bloom in autumn, it can be a tossup to decide whether these unique shrubs are winter-blooming or spring-blooming. But whatever we call the season, these shrubs with their unusual, fragrant flowers are welcome in the garden.

The yellow, gold, orange, or red flowers have a unique look thanks to their twisted, narrow petals, which dangle like small bright ribbons from the bare branches. The diminutive blooms are only 0.5–0.75 inch (1.2–1.9 cm) in diameter, with four narrow, straplike petals. In spite of their small stature, the flowers create a heady scent.

Hamamelis that bloom in late winter/early spring include *H. vernalis*, vernal witch hazel (*vernal*, "of spring"); *H. mollis*, Chinese witch hazel; *H. japonica*, Japanese witch hazel, and *H. × intermedia*, a group of many hybrid cultivars resulting from crosses between the Chinese and Japanese species. The name "witch hazel" probably originated from the early settlers' practice of using the forked branches of the witch hazels for divining rods.

All witch hazels are loosely spreading shrubs with an open habit. They generally grow from 6–15 feet (1.8–4.5 m) tall, depending on the type, and spread as wide. The foliage is reminiscent of hazelnut (hence, witch *hazel*), but less glossy. In the fall, the soft green leaves turn a beautiful yellow which is sometimes tinged with purple or red.

Cold-hardiness of witch hazels varies with species. All of the species discussed here are hardy to at least zone 5; *Hamamelis vernalis* is reliably hardy to zone 4. These plants have the interesting ability to curl up their flower petals when the temperature drops too low, a mechanism that protects the blooms from getting damaged by late cold snaps and adds to their value in the landscape. Witch hazels do best in moist soils and grow well in full sun or partial shade. There are no serious diseases or pests. The roots are sensitive to disturbance, so plants should be moved either in containers or balled-and-burlapped to avoid problems following transplanting.

Propagation of witch hazels is difficult and is probably best left to nursery professionals and serious horticulturists. Cuttings of very soft, succulent young growth are wounded slightly by scoring or scraping the cut end, then dipped in rooting promoter and rooted under mist; however, they may take in the range of three months to root, and a large percentage of the rooted cuttings will die. Seed requires two resting periods at different temperatures before it will germinate. Most cultivars are propagated by grafting onto seedling understock.

There is a plethora of good cultivars of the winter-blooming/spring-blooming witch hazels (*H. japonica*, *H. mollis*, *H. vernalis*, and *H. × intermedia*). One of the most widely grown and popular cultivars is 'Arnold Promise', from the Arnold Arboretum (Plate 26). It blooms in late winter with large, showy yellow flowers. 'Ruby Glow' (Plate 27) has bronzey to copper-red flowers and good red to orange fall color. Two other hybrid cultivars that are real standouts are 'Primavera', which blooms very prolifically in early spring with large, exceptionally sweet-scented, soft yellow flowers, and 'Sunburst', which flowers abundantly very early, in late January, with lovely but scentless lemon-yellow flowers.

Spring-blooming witch hazels make beautiful additions to the late winter/early spring landscape. Their delightful fragrance and interesting flower form are generally underutilized, yet so many sites would be transformed by the addition of just such a plant. The freshness of the witch hazel's fragrance helps us sense the possibilities of spring at its very start.

Ilex 'Nellie R. Stevens' / Nellie R. Stevens holly   PLATE 28

Ilex 'Nellie R. Stevens' combines the beauty of English holly with a decided tolerance for summer heat and humidity, a valuable attribute in hot-summer regions where English hollies languish. A hybrid of English holly (*Ilex aquifolium*) and the heat-tolerant Chinese holly (*I. cornuta*), 'Nellie R. Stevens' combines classic beauty with adaptability. As a bonus, this holly has the excellent vigor of most hybrids and grows much more rapidly than many other hollies. In good locations, it can grow 2–3 feet (0.6–0.9 m) in one season.

Nellie R. Stevens holly is a broadly pyramidal tree that can grow 30–40 feet (9–12.4 m) tall with great age. The foliage is typical holly: a glossy, deep green that stays dense and beautiful year-round. This hybrid bears bright red fruit very prolifically. Unlike many other hollies, 'Nellie R. Stevens' does not require a male pollinator to produce at least some fruit. 'Nellie R. Stevens' is an unusual female cultivar that can produce some red fruit without a male tree nearby.

Nellie R. Stevens holly is hardy to zone 6. It is exceptionally tolerant of heavy, wet soils and high temperatures, and is also drought tolerant. It performs best in full sun but will tolerate partial shade. For best growth, give the plant the best drainage available, and avoid really wet sites. 'Nellie R. Stevens' is tough and durable, but it can be susceptible to some scale problems. It can be propagated from stem cuttings taken almost any time of the year and rooted under mist. Quality plants are readily available at many nurseries.

Because of its rapid growth and year-round beauty, Nellie R. Stevens holly is a very good choice for both massed and specimen plantings. Planted 3–5 feet (0.9–1.5 m) apart and left unpruned, a row of these hollies with their dense, rich green foliage quickly grows into a perfect hedge, screen, or backdrop for colorful perennials. For use as a single specimen tree, selectively remove the lower limbs as the plant grows to create a single- or multiple-trunk small tree specimen.

At the North Carolina State University Arboretum, a tall, arresting hedge of darkly glistening Nellie R. Stevens hollies serves as the backdrop for the beautiful perennial border. The hollies form a divider between the perennial border and the white garden, on the east side of the Arboretum. Fast-growing Nellie R. Stevens holly is a useful plant in any garden. It is especially valuable in southern landscapes, where its darkly glistening foliage stays fresh even through a grueling summer.

Illicium spp. / Anise tree   PLATE 29

Low-maintenance broad-leaved evergreens of moderate stature, with seasonal interest and no disease or pest problems, are plants at a premium among gardeners, landscapers, and nursery professionals. This is especially the case in the hot, humid areas of the southeastern United States where the classic _Rhododendron_ is intolerant of clay soils and high temperatures, and where the long-reliable redtip, _Photinia_ × _fraseri_, is in serious decline from a blight epidemic. In the search for an alternative medium-sized, broad-leaved evergreen, the genus _Illicium_ shines as a promising constellation of useful garden stars.

Illicium, anise tree, is a genus of evergreen shrubs or small trees with species native to the southeastern United States, Japan, and China. Of varying habit and hardiness, these tough shrubs are characterized by neat olive-green foliage and interesting star-shaped, waxy flowers that bloom throughout the growing season. They are versatile evergreens for use in masses or as specimens. The general lore is that _Illicium_ requires at least partial shade and moist, well-drained soils to perform well. The reality, how-ever, is that once a plant has established a good root system after transplanting, it will thrive and grow in full sun, presenting a more compact shape and showing increased flowering. Foliage will yellow slightly at first in full sun due to moisture stress but will then green up as the plant adapts to the site. A number of _Illicium_ species perform beautifully in heavy clay soil. In more northerly areas, there is concentrated bloom in spring. In the Southeast, however, all _Illicium_ flower heavily in the spring or early sum-mer, and two species, _I. floridanum_ and _I. mexicanum_, also bloom sporadically through-out the summer and lightly again in fall. Anise tree gets its name from the distinctive fragrance of its foliage. When leaves are broken or brushed, the strong scent is remi-niscent of anise.

Illicium are not subject to any pest or disease problems. The plants are readily prop-agated from fresh seed, or from semihardwood cuttings collected in late summer and rooted under mist. The different species offer varying character in terms of flower size and color, cold hardiness, and plant habit.

Perhaps the most commonly available species is _Illicium parviflorum_, small anise tree, a native of the lower southeastern United States that can be found growing in wet areas from Georgia to Florida. It is informally pyramidal and upright and will ulti-mately reach 10–15 feet (3.1–4.5 m), although it is usually seen at heights of 6–8 feet (1.8–2.4 m) in the landscape. Foliage is a lighter olive than some other species, and the small 0.5-inch (12.7-mm), creamy yellow flowers, borne in early summer, are some of the least showy of the genus. This is probably the toughest landscape species. It thrives in sun and shade, and in heavy clay soils and dryer sites. It is reliably cold-hardy to zone 6.

Another readily available species is _Illicium floridanum_, often confused with _I. parv-iflorum_ in the trade. Florida anise tree is native to Florida but hardy to zone 7. Ironically, it differs from small anise tree in being smaller, reaching 8 to 10 feet (2.4 to 3.1 m) high in nature although generally seen at 5–6 feet (1.5–1.8 m) in the landscape, with a more compact habit. It also differs in flower color and size. Florida anise tree's flowers are much larger, 2 inches (5 cm) across, and usually dark wine in color (red, pink, and whitish forms have been reported in the wild). The flowers look like dark, emaciated stars fallen among the foliage and are an interesting feature in the spring. A white-flowered cultivar called 'Alba' has creamy white blooms, and the cultivar 'Halley's Comet' has brighter red flowers than the species.

There are some other interesting species that you may encounter. _Illicium anisatum_, Japanese anise tree, native to lower Japan and China, is reliably hardy only to zone 8. It will reach 10 feet (3.1 m) in height with a formally pyramidal habit. Flowers are light green fading to cream, 1–2 inches (2.5–5 cm) across, and with many more petals than other species, giving them the look of pale sea anemones. 'Pink Stars' is an exceptional seedling selection of this species, with extraordinary large and numerous blush-pink blooms. This cultivar, from the North Carolina State University Arboretum, Raleigh, North Carolina, is now in production by a few nurseries and should become more available with time. *Illicium henryi*, Henry anise tree, is native to China and likely hardy to zone 7. It is an 8-foot (2.4-m) shrub with especially dark green foliage and 2-inch (5-cm) flowers ranging from pink to maroon in color. *Illicium mexicanum*, Mexican anise tree, is native to Mexico. It offers cinnamon-red 2-inch (5 cm) flowers and good dark green foliage, and appears to be hardy to at least zone 7.

The broad-leaved evergreen anise trees deserve much wider use. Their interesting flowers, handsome foliage, and reliable performance in the landscape make them excellent alternatives to the usual evergreen shrubs.

Juniperus horizontalis / Groundcover junipers   PLATE 30

A living mosaic of greens, golds, blues and grays, with fascinating texture twelve months of the year, can be yours with the many cultivars of *Juniperus horizontalis*, groundcover juniper. This low-spreading, coniferous evergreen is a standard groundcover in the nursery trade across the United States. A few cultivars with excellent horticultural traits (such as 'Blue Rug'/'Wiltonii' and 'Bar Harbor') are widely grown, but more than fifty other cultivars exist. Some of these less frequently cultivated selections offer a range of height, color and texture that holds great potential for gardeners in North America. The foliage of *Juniperus horizontalis* cultivars is attractive year-round, with many showing ornamental changes in winter color.

Groundcover junipers are excellent choices for the tough conditions of new, urban landscapes. They thrive only in full sun but tolerate a range of soils from heavy clay to dry sand, and they flourish in low-maintenance environments. They are completely hardy to at least zone 3. These shrubs are readily propagated by hardwood cuttings in both winter and late summer. Some junipers are susceptible to foliar diseases in the cool, wet periods of the year. *Juniperus horizontalis* exhibits a range of resistance to this problem; all of the cultivars described here showed little or no disease susceptibility in my experience with woody plants at the North Carolina State University Arboretum in Raleigh, North Carolina.

At the Arboretum, master's thesis student Laurence Hatch, under the direction of Dr. Paul Fantz, assembled a reference collection of over 80 different *Juniperus horizontalis* selections for study and evaluation. These unusual varieties exhibit diverse characteristics of color, texture, height, and shape—welcome additions to the palette of groundcovers for landscape use. The foliage of *Juniperus horizontalis* cultivars is attractive twelve months out of the year, with many showing appealing changes in winter color—an excellent trait for winter interest in the landscape.

Following is a sampling of some of the best of these lesser known cultivars. All have performed well at the Arboretum.

'Argenteus' is a rich blue-green form with new growth in the spring emerging a contrasting gray. It reaches 11 inches (27.8 cm) in height with somewhat upright, cordlike branches and a 5-ft (1.5-m) spread.

'Blue Horizon' is an exceptionally flat form with intensely blue foliage. It has been one of the best cultivars at the Arboretum. The foliage, which is of excellent quality, turns an attractive bronze-green in winter. With a height of only 7 inches (17.7 cm) and a spread of 5 feet (1.5 m), 'Blue Horizon' remains uniformly flat without mounding around the central crown—a useful trait for large plantings.

'Douglasii' (or 'Waukegan') is a rapid grower with upright fans of blue-gray foliage on trailing branches, an especially nice texture for banks. Particularly tolerant of light, dry soils, it has also thrived in clay. It reaches a height of 17 inches (42.5 cm) and spread of 6 feet (1.8 m).

'Heidi' is a very slow-spreading form, reaching 15 inches (37.5 cm) in height and 3 feet (0.9 m) in spread, with unique, fernlike foliage in a unique gray-green color. It should be placed where the attractive fernlike foliage is seen from close proximity.

'Jade River' is a form with beautiful blue-green foliage with a slight silvery cast on long, spreading branches arching over a low and tight crown. It reaches 8–10 inches (20.3–25 cm) and spreads to 6 feet (1.8 m) with a softly undulating appearance.

'Lime Glow' is a spectacular selection discovered by Laurence Hatch in Raleigh, North Carolina, as a branch sport of 'Andorra Compacta'. The habit is upright and mounded, and the foliage is a bright lime-green to lemon-yellow. Height is 16–20 inches (40–50 cm) with a spread of 2 feet (0.6 m).

'Mother Lode' is relatively slow growing, reaching 12 inches (30 cm) in probable maximum height, with unbelievably bright gold, plume-textured foliage. It is an unusual ground cover for winter interest as well as a unique specimen plant. Plants at the Arboretum are younger than others in the collection and have reached 8-10 inches in height with a 2-foot (0.6-m) spread at about 3 years; eventually, the plant should reach a height of 10–12 inches (25–30 cm). This is the only patented cultivar listed here, and licensing from Iseli Nurseries in Oregon is required to grow this selection, which Jean Iseli discovered as a branch sport of 'Wiltonii'.

'Prince of Wales' is a rapidly spreading, exceptionally cold-hardy selection from Canada with extremely tight, low, bright green foliage that acquires a pronounced blue-purple bloom in the winter. It offers very good foliage quality and tight, dense form, reaching only 12 inches (30 cm) in height with a spread of 7 feet (2.1 m).

'Turquoise Spreader' is a cultivar from Monrovia Nursery that creates a wide-spreading plant of 12-inch-high (30 cm), rich turquoise-green, feathery foliage on long branches to 6 feet (1.8 m). Foliage quality is excellent and is tinged with gray-purple in winter.

'Variegata' is a low-spreading blue-green form with creamy new growth. Branches are a bit more upright than other cultivars, which adds to the impact of the variegation. It is effective combined with other full-sun groundcovers, including other groundcover juniper cultivars.

'Watnong' is a very low, tight form with clear green, beadlike foliage to a height of only 6 inches (15.2 cm). Excellent, especially dense groundcover capability with beautiful foliage spreading to 5 feet (1.5 m).

'Yukon Belle' is an extremely hardy form (to zone 2) with bright blue, somewhat rough textured foliage with plumelike branches, to 13 inches (32.5 cm) in height. The dense foliage creates an excellent blue mat spreading vigorously to 8 feet (2.4 m). The silvery tinge of the blue summer foliage changes to a darkened purple-green in winter.

These cultivars are only a small fraction of the many excellent *Juniperus horizontalis* selections. All of the cultivars listed offer unique ornamental characters in combination with good adaptability to landscape conditions. The range of spread and height characters described here reflect the important effect of time and climate on growth. Spread in particular continues to change over the life of individual plants, to potential sizes beyond those given for young plants.

Groundcovers are often relegated to monoclonal plantings that simply fill space that is not physically appropriate for lawn. The lesser known cultivars of *Juniperus horizontalis* can be used to create groundcover plantings that not only cover the ground, but also bring a whole new dimension to the landscape. Any number of unusual cultivars of groundcover juniper can be grown woven together to create a living carpet of pattern in the landscape. The soft gold of 'Mother Lode' curling through the silvery foliage of 'Jade River', or the play of textures when 'Watnong' and 'Heidi' intertwine, cannot be duplicated by any other type of planting.

Juniperus squamata / Singleseed juniper   PLATE 31

Singleseed juniper is a cool blue shrub that holds its soothing color through all four seasons. It is a native of the Himalayan mountain area and adjacent areas of western China, but the wild species is almost never seen in cultivation. A number of readily available cultivars offer a wide range of forms, but chief among their attractions is their excellent bright blue foliage, which retains its color well even through the hottest summer.

The foliage of singleseed juniper is generally short-needled, and more coarse and prickly in texture than that of the more commonly grown Pfitzer-type junipers (which are wide-spreading plants with arching branching, reaching about 5–6 feet [1.5–1.8 m] tall with twice the spread, and having fine, awl-like needles held tightly pressed to the stem), but texture and appearance of the foliage varies somewhat among the cultivars. The growth rate is moderate, making it a good choice for small-scale urban and suburban gardens.

Singleseed juniper is completely hardy to zone 4, and it is quite tolerant of drought and tough, urban sites. Full sun is important for good growth and foliage quality. Bagworms will take to this juniper if there is already significant infestation of the area, but *Juniperus squamata* is usually not a prime target of this pest (as Leyland cypress are). Plants can be successfully propagated from cuttings taken in late November through late February, then treated with moderate concentrations of rooting hormone, and rooted under mist. Alcohol-based rooting hormones may burn the cuttings.

Singleseed juniper has previously had a reputation for performing poorly in areas with hot, wet summers, but at the North Carolina State University Arboretum, selected cultivars have performed extremely well in the heavy, wet clay soils and high night temperature conditions of the Piedmont. They have been excellent, reliable sources of good blue color year-round.

Noteworthy cultivars include 'Blue Alps', an excellent upright form with a loosely pyramidal shape. The branchlets nod away from the main stem, giving it a nicely informal outline. It reaches 5–6 feet (1.5–1.8 m) in height, with spread of about 2–3 feet (0.6–0.9 m). Foliage is a sprightly silver-blue that, in my experience, remains in perfect condition year-round. This is a good choice for vertical interest in a small informal garden or small-scale mixed border.

'Blue Carpet' is a gray-blue spreading groundcover with a coarser texture than the *Juniperus horizontalis* types more commonly used as coniferous groundcovers. 'Blue Carpet' is also slower growing than its *J. horizontalis* cousins. It is a good choice for almost any site in full sun.

'Blue Star' (Plate 31) is probably the most commonly available cultivar at this time. This is a very low, mounding form eventually reaching 3 feet (1.8 m) by 3 feet (1.8 m) in spread, with an eventual height of 2–3 feet (0.6–0.9 m). It is a brilliant blue color all year and has the added attraction of a softer texture of foliage because of the more juvenile type foliage, which remains denser and closer to the stem. It combines beautifully with low perennials and bulbs. It is also a delightful container plant. Beware of letting mixed plantings cast too much shade on this plant or it will spread out and get ratty-looking.

'Holger' is the most Pfitzer-like form of the cultivars described here, showing the loosely arching, upright-spreading form and finer needle texture of a Pfitzer-type juniper, *Juniperus chinensis* ('Holger' may actually be a hybrid between *J. squamata* and *J. chinensis*). 'Holger' has a boxier shape than other forms, reaching 4–5 feet (1.2–1.5 m) in height with equal spread. The foliage is a bright blue-gray color in winter changing to a sprightly blue-green in summer. New spring growth emerges gold-yellow and frosts the branches like a dusting of gold powder—a beautiful effect in contrast to the blue foliage. 'Holger' is an attractive and reliable Pfitzer alternative, but it is somewhat more slow growing than the Pfitzers.

'Meyeri' is the parent of 'Blue Star' and 'Blue Carpet', but that's where the relationship ends. Its coarse, prickly foliage is a fabulous electric-blue held on upright, spreading branches. The form of this plant is loosely rounded and upright. It reaches 6 feet (1.8 m) in height with a spread of 4–5 feet (1.2–1.5 m). The foliage usually dies back and thins out on the lower part of the branches with age, so it does not remain attractive as a mature specimen.

'Prostrata' is a very slow growing, low-spreading form with gray-green foliage. It is very difficult to distinguish from *Juniperus procumbens* 'Nana', the well-known Japanese garden juniper, and is frequently confused with this other popular juniper in the trade.

The cultivars of *Juniperus squamata* offer bright, reliable color all year long and combine easily with other, shorter-lived garden stars. These versatile and adaptable junipers add valuable texture and a cool blue hue to the landscape, while remaining well-behaved in combination plantings.

Loropetalum chinense / Chinese loropetalum PLATES 32, 33

The beauty of freshly fallen snow is difficult to capture in words. In the garden, it drapes branches and foliage with a soft white coat, punctuated here and there with deep green leaves. That same artistry comes alive in *Loropetalum chinense*, a delightful evergreen shrub that covers itself in spring with a drift of snowy white flowers.

Chinese loropetalum, as the name suggests, is native to China (as well as a very limited area of Japan). It was introduced in the late 1800s, when many plants from Asia first appeared in American gardens. It is a large shrub normally reaching 6–10 feet (1.8–3.1 m) in height (older plants occasionally reach 20 feet [6.2 m]), with an equal or greater spread. Branching is somewhat irregular, giving the shrub an interesting undulating profile. The glossy evergreen foliage is dark emerald-green. The small, neat leaves are 1–2 inches (2.5–5 cm) long and somewhat rounded. The foliage, which stays attractive all year, creates a rather dense surface of leaves that follows the outline of the plant and contributes to its interesting form. The bark of this shrub is a warm nut-brown color, but it is generally hidden by the foliage.

Chinese loropetalum is most ornamental when in flower in early to mid-spring. It is a relative of witch hazel (*Hamamelis* spp.) and has similar flowers, with each bloom being a cluster of several very narrow, strap-shaped petals. The flowers have a sweet fragrance. Flowering is very prolific, with blossoms tightly clustered along the branches. This density of bloom creates the "blanket of snow" effect over the surface of the entire shrub.

Chinese loropetalum is hardy to zone 8 and can be grown in protected sites in the warmer areas of zone 7. It thrives in moist, well-drained, acidic soils, and also performs well in clay, but it is not very tolerant of dry soils. With adequate moisture, this versatile shrub will thrive in many different sites.

Full sun will give the most prolific flowering, but Chinese loropetalum also does quite well in significant shade, a very helpful trait in gardens with large trees that cast essentially constant, moderate shade. There are no disease or pest problems. Chinese loropetalum propagates readily from softwood cuttings taken in summer, treated with rooting promoter, and rooted under mist.

In addition to the white-flowered form of this shrub, some rare magenta- and pink-flowered forms have entered the United States in the last several years. The North Carolina State University Arboretum, the Atlanta Botanical Garden, the Arnold Arboretum, and the U. S. National Arboretum have all worked with various clones of these plants. The National Arboretum has released two named selections: 'Burgundy', with bold electric-magenta flowers and bronze-purple foliage; and 'Blush', with green foliage and lighter pink flowers. It is possible that the pink-flowered forms will be less hardy than the white-flowered plants; testing will be required to learn where they will be useful landscape plants.

Chinese loropetalum makes an excellent specimen plant or massed grouping. While it requires little or no pruning, it can be limbed up to make a small, multitrunked tree.

The beauty of Chinese loropetalum in full bloom brings a hint of winter's snowy artistry to gardens where real drifts are rare. At the North Carolina State University Arboretum, a large Chinese loropetalum brings great drifts of floral snow to the white garden each spring. By planting this lovely shrub in your own garden, you can add the quiet feel of a blanket of snow—without paying the price of frost.

Magnolia denudata / Yulan magnolia   PLATE 34

There is an elusively exotic character that all magnolias bring to the garden, whether deciduous or evergreen. Their creamy, fragrant flowers unfold as if some garden magician had persuaded pastel waterlilies to flower on the branches of a tree.

Early spring is the time when deciduous magnolias are in their glory in the garden. Most commonly seen are the well-named star and saucer magnolias. The flowers of star magnolia do indeed resemble a constellation of stars fallen among its light grey bark while the large blooms of saucer magnolia sit balanced along its branches like hundreds of lovely, floral saucers. While these two magnolias are very beautiful, they are not the only garden-worthy deciduous magnolias. There are many, many other species and hybrids that bring their own special character to spring. One of the most beguiling of these lesser known deciduous magnolias is *Magnolia denudata*. Yulan magnolia, named for its region of origin in central China, has been a favorite subject of Chinese artists for hundreds of years.

Named by the famous father of botany, Linnaeus, for a French botanist, Pierre Magnol, magnolias are an interesting and ancient group. They are the oldest of the true flowering plants and, though not very sophisticated biologically, they are very successful on a global basis. They have also been, understandably, very popular with horticulturists (both ancient and modern), and many special forms have been selected and hybridized.

Magnolia denudata is a small- to medium-sized tree reaching approximately 35 feet (10.9 m) in height. Eventually the crown of the tree will open to give a spread almost equal to its height, but a young tree stays upright with a tight habit, similar to young saucer magnolias. The flowers of Yulan magnolia are a magnificent 6–7 inches (15.2–17.7 cm) wide, with usually nine broad, waxy petals. Blooms of the species are a creamy white color and hold their upright, cupped shape longer than most of the saucer magnolias. The scent of Yulan magnolia flowers combines the heady sweetness of *Magnolia grandiflora* with the lemony character of the star magnolia and adds its own earthier notes of rich spice. There is no fragrance in the garden that can match that of Yulan magnolia in early spring.

The leaves of *Magnolia denudata* are an attractive medium green with an interesting shape. They are widest near the tip, curving into a sort of pear shape with a pointed end. The bark of the tree is somewhat darker than that of other deciduous magnolias, and it makes an effective foil for the light-colored flowers, which open before the leaves in late February or early March.

Like other magnolias, Yulan magnolia is adaptable to a wide range of landscape conditions. It is cold-hardy to zone 5, but it is an early bloomer—one of the earliest magnolias to flower—and so the buds and blossoms can be damaged by frosts and freezes at the end of winter. *Magnolia denudata* grows best in evenly moist, somewhat acidic soils that are high in organic matter, but it also performs well in clay soils. Roots are fleshy and tender, so the trees should be transplanted balled and burlapped or from a container, but once established this is a strong tree that is quite tolerant of urban stresses, including air pollution. Full sun will give the best flowering, but the tree will also remain healthy in light shade.

In addition to the species with its cream-colored flowers, there are other color forms that can be found with some determined hunting. These colored forms are no longer considered to be valid *Magnolia denudata* by botanists, but they will be found labeled as such in the trade. The botanical variety *Magnolia denudata* var. *purpurescens* has flowers with a deep rose exterior and light pink interior. The flowers of the cultivar 'Purple Eye' are white with a lavender base; those of 'Forest's Pink' are an attractive pink. 'Swada' has white flowers on a compact plant. 'Wada' bears small white blooms that open later than the species. Vegetative propagation of Yulan magnolia is difficult, and named selections are generally grafted onto saucer magnolia rootstock.

The North Carolina State University Arboretum in Raleigh, North Carolina, includes one of the largest collections of deciduous magnolias in the United States. It is an incomparable sight to walk through the magnolia collection in early spring when thousands upon thousands of magnolia blossoms fill the air with their color and fragrance. But even among this multitude, the flowers of *Magnolia denudata* are unforgettable as they rest on their branches like perfect ivory bowls, filled with a fascinating, complex fragrance. Yulan magnolia is a treasure for any spring garden.

Magnolia stellata / Star magnolia   PLATES 35, 36

The deciduous magnolias are the earliest of the magnolias to flower, unfolding their blossoms to reveal a charming beauty that is all the more delightful as it often dares the worst of winter's last efforts. One of the very earliest to bloom is star magnolia, *Magnolia stellata*. *Stellata*, Latin for "starry," is a lovely choice of name because the flowers are indeed like white stars fallen to earth.

Magnolia stellata is a shrubby small tree with a dense, somewhat oval habit. It is multistemmed and, with maturity, can reach 10–20 feet (3.1–6.2 m) in height and spread to 10–15 feet (3.1–4.5 m) in width. In the very early spring, while there is still danger of frost, the flowers open into sweetly fragrant clusters of 10 to 20 narrow petals, each 2 inches (5 cm) long. The many long petals give the flower its starlike character. Most forms of star magnolia have white flowers with yellow stamens in the center that add a touch of gold. The starry flowers open before the leaves have emerged to distract from their shape.

When the leaves do come out, they mature to a pleasing grass-green and retain their quality appearance through the summer. Sometimes they develop a bit of bronze-yellow fall color but it is not usually significant. The spring bloom is star magnolia's big show for the year, but this little tree also has a certain appeal in winter. The branching habit and light gray bark are appealing, and the large, furry flower buds decorate the branch tips like fuzzy mittens.

Because they bloom early, deciduous magnolias like *Magnolia stellata* are often at the mercy of unpredictable spring weather. While the plant itself is completely hardy as far north as zone 4, the blossoms will suffer damage from a late spring freeze if they have started to open. For this reason, it is a good idea to plant any deciduous magnolias in protected sites; however, this can be overdone. If they are planted in a very warm site (such as against a south-facing wall), they may bloom even earlier and be more subject to frost damage than if they were in a cooler, shaded site. Probably the best solution to this balancing act is to plant deciduous magnolias in a site that is not exposed to winter winds, and, perhaps, to keep them away from the south side of buildings. The periodic disappointment in years that flowers are damaged will be more than offset by those incredible springs when the starry blossoms cover the tree like the Milky Way.

Aside from the late-freeze problem, *Magnolia stellata* is a trouble-free landscape plant. It is not a rapid grower, and therefore makes a good specimen plant for the home garden, where its scale is never overwhelming. Star magnolia flourishes in a loamy soil that is rich in organic matter, but it also does well in many other soils and is quite tolerant of summer heat and heavy soils.

Star magnolia can be propagated either from seed or vegetatively from cuttings. Seed is stored in a moist medium, such as peat moss, in plastic in the refrigerator for three months before sowing. Cuttings can be taken after the prominent flower buds have formed, usually in June through July. Cuttings are treated with rooting promoter and rooted under mist.

Star magnolias are readily available at garden centers. With some hunting through specialty nursery catalogs, you can also find a number of excellent cultivars. 'Centennial' was selected at the Arnold Arboretum in Massachusetts to commemorate their hundred-year anniversary. Its lovely white flowers, tinged with pink on the outsides of the petals, are held more widely open than those of the species. 'Chrysanthemumiflora' has an awful name but a fabulous flower, with an extraordinary number of light pink petals per bloom. 'Rosea' boasts pink flower buds, which open into pink flowers that fade to white as they mature. 'Royal Star' is one of the best known and most readily available cultivars. It is very fragrant and has perhaps the most petals per bloom. Its flowers open from light pink buds to a clear white. 'Rubra' (Plate 36) has large lavender-rose petals that open to pink. 'Waterlily' is aptly named, because the petals are narrower than most and the white flowers are reminiscent of waterlily blooms. Both 'Royal Star' and 'Waterlily' are later blooming than others and so may avoid some of the danger of frost damage.

Each year, star magnolia braves the uncertain early spring, greeting the first mild days with a show of sweetly fragrant flowers. Blooming far ahead of their showier cousins, the little star magnolias are even more welcome because of the precociousness of their starry beauty.

Magnolia virginiana / Sweetbay   PLATE 37

During early summer, a season when few trees and shrubs offer flowers or fragrance, this lovely North American native provides both. *Magnolia virginiana* is a deciduous-to-evergreen magnolia, depending on climate. Native to the southeastern United States, sweetbay blooms long after the early deciduous magnolias have finished their displays. In June, the glossy foliage is sprinkled with creamy white cupped flowers. The 2–3-inch (5–7.6-cm) blooms are sweetly fragrant, with the scent of lemons, yet are not overpowering.

Sweetbay ranges in habit from a small, multistemmed deciduous tree in the North to a large evergreen, pyramidal tree in the warmer South. Depending on where the garden is, *M. virginiana* will generally reach 20 feet (6.2 m), with half the spread, as a medium-sized tree. It can reach 40 feet (12 m) with age in the South. In the Piedmont area, it usually reaches 20–30 feet (6.2–9.0 m) with a spread of 10 feet (3.1 m) and retains a tree-type habit.

Sweetbay has beautiful foliage as well as lovely flowers. The light gray-green leaves, 3–5 inches (7.6–12.7 cm) long and only 2 inches (5 cm) wide, are much narrower and somewhat shorter than those of most evergreen magnolias. The bottoms of the leaves are a beautiful silvery color that is revealed when a breeze blows through the tree, creating a silver wave among the branches. *Magnolia virginiana* also produces attractive fruits, which are similar to all magnolia fruits: a spiraled, upright conelike cluster of seeds, which are usually colored bright red. The seed of sweetbay's fruit is bright carmine in the fall.

Sweetbay is cold-hardy to zone 5. It is evergreen throughout areas with mild winters, including much of the Southeast, but is semievergreen to deciduous in colder regions. In the wild, sweetbay grows in wet, acid lowlands. As a landscape plant, it grows best in acid soil, without which it may be subject to chlorosis. It is tolerant of wet, shaded sites in the landscape, a unique and valuable attribute for a magnolia. Sweetbay suffers from no serious diseases or pests. It is not difficult to propagate from summer softwood cuttings taken from young (two- to three-year-old) plants, treated with rooting hormone, and rooted under mist. Cuttings from mature trees may be more difficult to root. Seed needs cold treatment for three to five months for good germination.

There are not many cultivars of *Magnolia virginiana*, but there has been some selection work, particularly for hardiness. 'Henry Hicks' is an especially hardy cultivar that remains evergreen at much colder temperatures than the species, but it has no other special ornamental attributes. The botanical variety *Magnolia virginiana* var. *australis* may be the plant of the southern range of sweetbay. Its branches have more pubescence than those of the species, and the plant is always evergreen, larger, and more consistently treelike than the species.

Magnolia virginiana is a trouble-free specimen tree for the smaller scale home landscape. Its glossy green, refined foliage is a restful foil for the subtle flower display. Sweetbay combines well with other, more showy and colorful woody plants and adds interest and beauty throughout the season.

Trees propagated from stock originally collected in Florida will not be as hardy as trees propagated from plants derived from stock collected in Virginia or New York, so it is a good idea to ask your nursery owner the "provenance," or geographic origin, of his or her propagation stock. Look for trees propagated from plants that originated in an area with a winter climate similar to that of the intended planting site.

Sweetbay is indeed a sweet tree that brings unassuming charm to summer gardens. To get an idea of the quiet beauty it can add to the garden, look for the creamy blooms of sweetbay as you drive past the streams and wet fields of southern country.

Pinckneya pubens / Poinsettia tree   PLATE 38

As native plants are rediscovered by horticulturists for use in the landscape, it is important to remember that these plants are not necessarily easy to grow. Some native plants are adapted to very specific conditions and need even more specialized care than "exotic" landscape plants. Yet these plants offer unique and beautiful qualities that are their own reward for any extra work they may require.

A good example of such a native is *Pinckneya pubens*, poinsettia tree, also known as hardy fever tree. *Pinckneya* is a small tree with glossy green foliage and an exotic flowering habit. It produces showy clusters of salmon-pink sepals reminiscent of poinsettia flowers. Originally named for its supposed medicinal qualities, this tree is a rare native of the coastal plain of the southeastern United States.

Pinckneya is hardy to zone 8 and protected areas of zone 7. The tree has a rather open habit and will reach 20–30 feet (6.2–9 m) high. It requires moist and well-drained soil, and thrives only when its roots are cool and shaded. It is easily propagated from seed which requires no pretreatment or from stem cuttings taken in early summer and rooted under mist.

Although it can be finicky about its growing conditions, *Pinckneya* is well worth the effort to have in the landscape. While it is a quiet plant for most of the year with little to offer in the way of fall color or winter character, *Pinckneya pubens* is spectacular in the spring, when the pink poinsettia-like sepals cover the crown of the tree like a flock of rare tropical birds blown up from the islands by a strange turn of the weather.

Prunus glandulosa / Dwarf flowering almond   PLATE 39

When the gardening bug hits full strength, none of us—novice or expert, estate designer or suburban home owner—can resist the call of the garden. Although sophisticated and costly plants may tempt us during this season of inspired gardening, there is no reason to feel that these are the only alternatives. Many affordable plants are available in local gardening centers and nurseries that are lovely when given the right care and setting. One such thrifty choice is the dwarf flowering almond.

Dwarf flowering almond is not an almond (*Amygdalus communis*) at all, but a close relative of flowering cherries. It is native to China and Japan. This little deciduous shrub reaches 4–5 feet (1.2–1.5 m) in height and 4 feet (1.2 m) in width, assuming an open oval or mounded form as it matures. Its star season is spring, when sweet pink or white cherrylike blossoms cluster along the stems before the leaves are out. The summer foliage is light green and resembles miniature peach foliage. Both single- and double-flowered forms are available. Single-flowered cultivars sometimes produce small red fruits in the summer.

Dwarf flowering almond is hardy to zone 4. Culturally, it requires a bit more attention than some plants. While generally well adapted to most landscape conditions, it should be planted in a site with the best drainage, fertility, and light available in the garden to maximize flowering. Because of its multistemmed, spreading habit, it can get a bit scraggly. Periodic pruning will rejuvenate the shrub and increase flowering. Dwarf flowering almond is easily propagated from stem cuttings rooted under mist.

A number of cultivars offer different flower colors and single or double blossoms. 'Rosea Plena' (sometimes called 'Sinensis'), which has double pink flowers and foliage of a richer green than most, makes a bit more showy statement in the garden than others. 'Rosea' produces single pink flowers. 'Alba' has single pure white blossoms. 'Alba

Plena' (sometimes referred to as 'Alboplena') shows off double white blooms with a pinkish tinge to the outer petals (Plate 39). 'Lawrence' is a single-flowered form with pinkish white blossoms that are perhaps the earliest of all, attaining full bloom as early as March in zone 7.

This sweetly unassuming and readily affordable flowering shrub is a great choice for small-scale suburban yards and gardens. Widely available and economically priced, dwarf flowering almond is a good beginning for new homes that invariably need something to perk up foundation plantings.

Prunus × *incamp* 'Okame' / Okame cherry   PLATE 40

In early spring, as visitors drive into the North Carolina State University Arboretum, they are greeted with clouds of rosy pink blooms that float beside the entrance driveway. These gentle clouds of bloom belong to Okame cherry, a lovely small tree that greets early spring with flowers shaped like diminutive bells, scented with a hint of pleasing fragrance. As the flowers open, the branches are hung with hundreds of delicate bells, riffling with the slightest of breezes.

The graceful *Prunus* × *incamp* 'Okame' is a small tree with the beautiful polished wine-red bark of many other cherries. It reaches 15–20 feet (4.5–6.2 m) in the landscape, with refined branching and a rounded to oval crown. The bell-like shape of the individual flowers is uncommon among cherries, and the deep rose pedicels add to the color of the flower display. 'Okame' usually reaches peak bloom in late winter in warm areas or early spring in more northern areas, before the leaves emerge. The foliage of this lovely tree begins as spring green and then darkens to emerald as the season progresses, finally turning rich bronze to bright orange in the fall.

'Okame' is a hybrid of *Prunus campanulata* and *P. incisa*, developed in the late 1940s by the renowned English cherry breeder, Collingwood Ingram. The 'Okame' hybrid was developed to increase the hardiness of *P. campanulata*, a native of Japan and Taiwan known as the Taiwan or bell-flowered cherry. 'Okame' stands up better to severe weather and has the added advantage of hybrid vigor with somewhat more prolific flowering than its parents.

'Okame' is hardy to zone 6, but late freezes may damage flowers in years when winters have been mild and the bloom is very early. Like most cherries, this cultivar thrives in moist, well-drained sites, but it also performs well in heavy clay soils. Full sun gives the best flowering display.

It is important to keep any cherry tree healthy to minimize the disease and insect problems that beset all the members of this botanical group and also to maximize flowering each year. Adequate soil fertility, light, water, space, and air movement are all important to maintain the health of the tree. Light pruning may be necessary to increase light and should be done shortly after flowering. 'Okame' can be easily propagated by rooting softwood cuttings taken in early summer. The cuttings are treated with rooting promoters and rooted under mist.

Prunus × *incamp* 'Okame' has been one of the recipients of the Pennsylvania Horticultural Society Gold Medal Award, a prestigious award for plants of excellence for American gardens. One of the most beautiful sights awaiting any gardener in the spring is that of 'Okame' in full flower, its deep rose bells floating above the daffodils blooming in concert underneath.

Prunus persica var. *nucipersica* 'White Glory' /
'White Glory' weeping nectarine   PLATE 41

A waterfall of white arches to the ground from the branches of 'White Glory' weeping nectarine. Cascades of white flowers on downwardly sweeping branches sculpt a floral display that is hard to rival in the spring garden.

Nectarines are a botanical variety of the familiar peach, technically called *Prunus persica* var. *nucipersica*. The character that separates them from peaches is the "fuzz-less-ness" of their fruits. Two traits separate 'White Glory' from the other nectarines, which are grown for their delicious fruit: its cascading habit and its flowers. 'White Glory' is a weeping tree, with branches that arch over and cascade to the ground, much like the familiar weeping cherries but with a stiffer, slightly coarser habit. Its flowers are larger and showier than those of other nectarines, but they do not result in good quality fruit.

'White Glory' is a medium-sized deciduous tree reaching 10–15 feet (3.1–4.5 m) at maturity and spreading to 15 feet (4.5 m). The arching habit of its branches causes the inner branches to die out with time (from the extra shade), which creates a magical cove under the tree, the perfect place for a garden bench as the tree ages. Glossy white 2-inch (5-cm) flowers emerge before the foliage to create the spectacular display. In late March in zone 7, the branches are covered with single and semidouble flowers. Soft, light green leaves follow the blooms. They resemble peach leaves in appearance but are longer and somewhat more drooping. No appreciable fall color develops, but the foliage is attractive during the summer and may turn a light clear yellow in autumn in years with cold, crisp weather. In winter, the bare, weeping branches create an interesting silhouette in the garden.

Peaches and nectarines are natives of Asia, but the 'White Glory' weeping nectarine was selected by F. E. Correll in 1960 from a peach breeding program at the Sandhills Research Station in North Carolina. It was introduced and made available to the nursery industry by Dr. Dennis Werner of the North Carolina State University horticultural science department. The cultivar is propagated by grafting budwood of 'White Glory' to rootstock of common peach cultivars (so the top of the tree is 'White Glory', while the roots are the sturdier peach). This process requires skill and practice and makes the plant more valuable and a bit harder to find in the trade.

Culture of 'White Glory' is similar to peach. It is reliably cold-hardy to zone 5; however, the earliness of its bloom makes it subject to damage from late spring frosts. It needs full sun and the best-drained soil you can provide. It will however tolerate heavier soils if they are not extremely wet. Full sun is very important for good flowering. Pruning out the dead wood that accumulates on the interior portions of the tree is simple and will minimize twig litter. Flowers are produced on the previous year's wood, so live wood should be pruned after flowering has finished in spring.

In spite of the risk of late spring frost, the breathtaking beauty of 'White Glory' in most years is worth the occasional year when the flowers are damaged. Once seen in full bloom, this unusual tree can never be forgotten.

Prunus × *yedoensis* / Yoshino cherry   PLATE 42

Yoshino cherry generally blooms before the leaves emerge, with clusters of white to pink, lightly fragrant flowers. The texture of the blooms is like delicate paper, as if a thousand origami artists had set out to decorate the tree in pastel clouds. A hybrid of somewhat uncertain parentage, _Prunus × yedoensis_ was found in Japan, but it has become popular around the world as a beautiful flowering tree.

Yoshino cherry is a full-sized, deciduous tree of softly rounded habit reaching 40–50 feet (12.4–15.2 m) in height with a gracefully arching spread to the canopy, but there are a number of cultivars of varying shapes and sizes. Its spring flowers may be pristine white to pale pink, depending on the cultivar. The flowers mature into minuscule glossy, blue-black fruit. Summer foliage is a deep green, and fall color may be vivid golds and oranges.

Yoshino cherry is completely hardy to zone 5. It prefers moist, well-drained, light soils but also does well in clay. Full sun is best for flowering and fall color, but the tree will also do well in very light shade. Yoshino cherry is propagated from seed, which requires two months of cold pretreatment layered in a moist medium, or from softwood cuttings taken early in the season and rooted under mist.

There are a number of excellent named selections of this hybrid cherry. 'Afterglow' has warm pink flowers. 'Akebono' has soft pink, double to semidouble flowers and reaches 20–25 feet (6.2–7.5 m). 'Ivensii' is a weeping, white-flowered form. 'Pink Shell' is aptly named; its shell-pink flowers fade to pale pink. 'Snow Fountains' (or 'White Fountain') is a semi-weeping form with clear white blooms that reaches only 10–12 feet (3.1–3.7 m) in height with maturity. 'Shidare Yoshino' is the cultivar name for the botanical form _Prunus × yedoensis_ f. _perpendens_, the most common weeping form, which is generally referred to as the weeping Yoshino cherry. 'Shidare Yoshino' is a showy weeper with white to very pale pink blooms. 'Yoshino Pink Form' is similar to 'Pink Shell' but may bloom slightly later.

At the North Carolina State University Arboretum, Yoshino cherry drapes cascades of pink delight over the beds of daylilies adjacent to the parking area. It is a delicate star of the spring flowering extravaganza. Of the many forms of Yoshino cherry, one is certainly the right choice for your own landscape, where its reliable, delicate display adds a special charm to spring.

Salix spp. / Pussy willow   PLATE 43

Soft and furry to the touch, and curious to the eye, pussy willows add a playful note to the garden. They satisfy both the mischievous child and the winter-weary horticulturist in us all, with velvety silver buds held aloft on sinewy branches.

Pussy willow is the common name for a number of different species of small willows whose tiny flowers emerge in the spring on mitten-shaped clusters called, appropriately enough, catkins. The catkins, which appear in early spring before the leaves emerge, have a soft, silvery look, an effect created by the tiny gray, silky hairs attached to the little flowers in the catkins.

All of the pussy willows are easy to grow. They are all hardy to at least zone 5 and are easy to transplant. They are tolerant of a range of soil conditions but do best in moist areas in full sun and can be planted in a relatively wet area of a garden where little else will grow. The genus name _Salix_ is from the Celtic name for willow (_sal_, meaning "near," and _lis_, meaning "water"). All willows, including pussy willows, are easily propagated from hardwood or softwood cuttings at any time of year.

The most common pussy willow is goat willow, _Salix caprea_, which is also the largest. An erect, multistemmed tree, it can reach 12–15 feet (3.7–4.5 m) in both height and spread. Its chief landscape interest is the early spring display of fuzzy catkins and it can be useful as an inexpensive note of interest in a difficult site. It is hardy to zone 4.

Rosegold pussy willow, *Salix gracilistyla*, is a lower growing. loose and open relative that shows off charming pink or red-tinged, extra-long catkins every spring (Plate 43). The catkins of this willow come out earlier and are larger and showier than those of *Salix caprea*, and the plant has thinner, more refined branches. Rosegold pussy willow has lovely foliage of soft gray-blue. It reaches 8–10 feet (2.4–3.1 m) with a spread of 5–10 feet (1.5–3.1 m). It is hardy to zone 5.

Black pussy willow, *Salix melanostachys*, is the dark cousin of rosegold pussy willow with a somewhat coarse habit. It is sometimes classified as *S. gracistyla* var. *melanostachys*. Its branches and catkins are deeply colored. The catkins emerge while the stems are still in their winter coloration of wine-black. The catkins open purple-black, often with a magenta strip, and the catkins become sprinkled with the yellow of the pollen—much like a technicolor wooly bear caterpillar. Catkin color is quite variable among individuals, so buy plants in spring to look for those with the deepest colored catkins. Black pussy willow makes a striking addition to a late winter and early spring landscape.

Even before the catkins emerge, a mixed planting of rosegold and black pussy willows in a moist corner of the garden creates a beautiful winter weave of gold and deep wine stems. In the spring, the branches become tufted with the rosy silver and rich purple furred catkins, creating a delightful welcome for early forays into the garden.

Salix chaenomeloides is perhaps the most elegant and consistently showy of the pussy willows. The graceful arching stems and exceptionally large, rounded buds are a deep red color, which makes a lovely offering for the winter garden. The catkins are exceptionally large and very ornamental as their soft, clear silver contrasts with the carmine-colored stems in the spring.

Branches of all of the pussy willows can be easily forced into early bloom. Cut long branches in late winter and place them in water in a warm place. In one to three weeks, the fuzzy catkins will begin to peek out, like kittens from under a blanket. Who can resist their velvety softness? Pussy willow branches can also be dried for permanent use by hanging branches upside down in the dark when the catkins have just emerged. After drying, they can be arranged in many different settings. Rosegold and black pussy willows and *Salix chaenomeloides* all make good additions to flower arrangements. Their stem colors, charming catkins, and graceful lines are becoming more and more popular with floral designers looking for a unique blend of color, form, and texture for late winter and early spring displays.

Pussy willows have often been regarded as just a novelty, or as "kid's stuff" in the garden. This trio of sophisticated pussy willows offers delights of winter color and form that go beyond, but never leave behind, the whimsical catkins that gave pussy willow its name.

Styrax japonica / Japanese snowbell   PLATE 44

Ring in delightful spring bloom with the Oriental charm of *Styrax japonica*, Japanese snowbell. A native of China, Japan, and Korea, this small deciduous tree was introduced to European gardeners in the late 1800s. It bears clusters of delicate, clear white bells with soft yellow centers against early chartreuse foliage.

Japanese snowbell grows to 20–30 feet (6.2–9 m) in height with comparable spread and acquires a broad, rounded habit as it matures. In midspring, pendulous, bell-shaped flowers with yellow stamens dangle from the branches as the spring flush of leaves matures on the upper side of the branches. The subtly fragrant flowers are about

0.75–1 inch (1.9–2.5 cm) wide and cover the undersides of the branches like clouds of sleeping butterflies. As the flowers expand, their petals arch back slightly to reveal the center of the flowers. The foliage is an attractive bright green and is held above the flowers fluttering below. The flowers mature into gray fruits that are of some interest, but the leaves do not develop significant fall color. The smooth bark of Japanese snowbell is an appealing gray-brown and develops very fine, cinnamon-orange colored fissures as the plant matures.

Cultivars of Japanese snowbell are worth seeking in specialty nurseries. 'Carillon' (Plate 44) is a weeping form; its arching branches are lined with tiny white floral bells and rich green velvet leaves. 'Pink Chimes' has an upright habit and blooms with one of the most delicately lovely shades of light rose-pink found in gardens. A third selection, 'Emerald Pagoda', is not yet available in the trade but will be released from the North Carolina State University Arboretum evaluation program. This special plant was found by J. C. Raulston on the island of Sohuksan in Korea during the 1985 United States National Arboretum plant collection expedition. 'Emerald Pagoda' is an upright form with much larger leaves and flowers than the species or other cultivars. The flowering display of 'Emerald Pagoda' creates a fantastic effect, especially in plantings near sloped walkways and other areas where garden visitors can look up into the flowers.

Japanese snowbell is one of the best ornamental plants for zones 5 and warmer. Culture is not difficult. It does best in moist, well-drained, somewhat acid soils, but it is a very stress-tolerant plant and will perform well in many more difficult sites. It has flourished in heavy clay soils and flowered beautifully.

Propagation of cultivars is easy from summer softwood cuttings, which are rooted under mist, but the trees can also be propagated by grafting. Propagation of the species can be done from seed, but the seed exhibits a double dormancy, making it more difficult to attain good germination.

Japanese snowbell is a unique beauty for the spring garden, with its sweet, snow-white bells suspended from bright green angel wings against a frame of smooth gray branches. This entrancing tree rings in the spring season with special delight.

Syringa oblata var. *dilatata* / Korean early lilac   PLATE 45

Gardeners in mild winter, hot summer areas know all too well that the common lilac, *Syringa vulgaris*, can't take the heat. Yet it's still possible for gardeners in these areas to enjoy the grandmotherly, back-porch appeal of lilac, by planting *Syringa oblata* var. *dilatata*, an early lilac with new adaptability and antique appeal. The *dilatata* variety is quite well adapted to climates with hot summers, such as the southeastern and southwestern United States, yet it is very similar to the common lilac in appearance and fragrance.

Syringa oblata var. *dilatata* is a large shrub that reaches 10–12 feet (3.1–3.7 m). Closely related to common lilac, it has similar foliage and flowers. The foliage differs in that it emerges in the spring as a warm bronze-red. The flowers are in clusters similar to those of common lilac and are the traditional lavender color with a very sweet fragrance.

Although it is cold-hardy to zone 3 and thrives in cold-winter climates, Korean early lilac does not require a long cold period before it will flower, as common lilac does. This trait contributes to the fact that early lilac is the first lilac to bloom in the early spring. The earliness is a real treat, but it can also make the flowers subject to damage from late spring frosts, although the beauty of the flowers is well worth the occasional disappointment.

The *dilatata* variety of early lilac, a native of Korea, is the best of the *Syringa oblata* species for landscape use in hot climates. Like most lilacs, it does best with excellent drainage. The most prolific flowering will occur in full sun. One of the exceptional things about *Syringa oblata* var. *dilatata* is that it has attractive fall color, an unusual trait in lilacs. The leaves turn bronzed red to purple in the autumn, which brings extra interest to the fall garden.

All lilacs are difficult to propagate dependably from cuttings. Many lilacs are now propagated by tissue culture, which is a way of growing a whole new plant from a tiny, almost microscopic, cutting. Lilacs can also be grown from seed, which requires one to three months of storage, layered in moist peat at 40°F (4°C).

As the sun warms the earth and robins are flocking in the garden, the appeal of traditional lilac blooms with their delicious fragrance is irresistible. Why resist? With proper selection, gardens in almost any climate can be home to the allure of lilacs. With a little adventurous gardening, these lilacs will be delighting you year after year.

Viburnum carlesii / Koreanspice viburnum   PLATE 46

Fragrance has been one of the great pleasures of gardening since humanity's first efforts at horticulture. Plants of every size, shape, and color have been selected and cultured over hundreds of years to provide fragrances from richly exotic to lightly aromatic. While fragrance may have been forgotten for a time in the breeding and development of visually exciting plants, the importance of fragrance in the garden has seen a great resurgence of late as gardeners turn back to their gardens to experience the beauty of the landscape with all of the senses.

One group of plants that seems imbued with a wealth of fragrances of all types (in a full range from the sublime to the beastly!) is the genus *Viburnum*. It often seems that there is a viburnum to fit every landscape niche and requirement. One of the exceptional species for spring fragrance is *Viburnum carlesii*, Koreanspice viburnum.

Koreanspice viburnum is native to Korea, and the scent of its flowers is a marvelous blend of spicy aromas. This deciduous shrub reaches 5–6 feet (1.5–1.8 m) in height, with an equal spread. It has an irregularly rounded, relatively dense habit and is somewhat slow-growing. The flowers develop before and as the leaves are first emerging in the spring. Tightly rounded clusters of pastel pink buds open to reveal small, slightly waxy white flowers that retain a pink tint on the outer side of the petals, giving the flower clusters a delicately pink cast. The flower clusters covering the plant are very showy in themselves, but combined with their hypnotic fragrance, they create an incomparable spring delight. Flowers mature into small, rather inconspicuous, bluish black fruit. The broadly oval, somewhat fuzzy foliage is an attractive gray-green. Leaves may turn burgundy in the fall, but they do not usually develop consistently good fall color.

Koreanspice viburnum is completely hardy to zone 4. It performs best when moved as a balled-and-burlapped young plant. The shrub grows best in moist, well-drained, acid soil, in full sun or partial shade, but it will also tolerate clay. Bacterial leaf spot and powdery mildew may be problematic on stressed plants.

Viburnum carlesii is often propagated by grafting scionwood onto rootstock of *Viburnum lantana*, which can lead to problems in the landscape because the rootstock is likely to sucker and overgrow the desired plant. Purchasing Koreanspice viburnum on its own roots will avoid this potential difficulty. Koreanspice viburnum is difficult to propagate vegetatively (hence the grafting approach) but, according to woody-plant expert Michael Dirr, it can be rooted successfully from cuttings taken in early summer and treated with high concentrations of rooting hormone.

A number of good cultivars of Koreanspice viburnum are available, as well as excellent named hybrids that have Koreanspice viburnum as one of the parents. Some of the cultivars of the species *Viburnum carlesii* include 'Aurora', with very deep red buds; 'Carlotta', with especially broad leaves; 'Compactum', a very dense, slow-growing form and a good choice for very small, walled gardens or near patios where the fragrance can really be appreciated; and 'Diana', which is especially vigorous and strongly fragrant.

Some gardenworthy hybrids with good resistance to bacterial leaf spot and powdery mildew are 'Cayuga', with particularly abundant clusters of flowers; 'Eskimo', a United States National Arboretum release of very dense habit with creamy white flowers (Plate 46); 'Mohawk' (a complex, backcrossed hybrid of the *Viburnum* × *burkwoodii* group), which combines the excellent fragrance and floral display of *V. carlesii* with brilliant scarlet to orange fall color; and *V.* × *juddii* (Judd viburnum), a fuller, larger plant than the species, reaching 6–8 feet, with larger flower clusters and allegedly reduced fragrance, although this is only evident to those with very sensitive noses.

The newer named hybrids with *Viburnum carlesii* as a parent may be superior to the species in terms of flowering habit and disease resistance, but they owe their beautiful fragrance to the influence of Koreanspice viburnum. When the various cultivars and related forms of *Viburnum carlesii* bloom, they never fail to attract visitors in buzzing droves, like bees to honey. Their irresistible fragrance is an incomparable addition to any spring garden, whichever cultivar you plant.

Viburnum tinus / Laurustinus viburnum PLATE 47

Viburnums are a vast group of flowering and fruiting shrubs that fill many needs in the landscape. The numerous species have been cultivated in so many forms that there truly are viburnums for all seasons. A less well known viburnum that is equally worthy of landscape attention for its year-round interest is laurustinus viburnum, *Viburnum tinus*.

Laurustinus viburnum is a beautiful, refined, broad leaved evergreen shrub that reaches 6–12 feet (1.8–3.7 m) and spreads somewhat less than that. It grows into a rounded upright form and remains dense with lustrous, dark green leaves. The foliage is very beautiful, evenly colored and glossy, like the coat on a well-bred horse. Individual leaves are somewhat reminiscent of those of the Carolina cherry laurel (*Prunus caroliniana*), but smaller.

The clusters of flowers are borne in the late winter and early spring. The buds expand as tiny pink punctuations that contrast beautifully with the dark evergreen foliage. The flowers open into pink-tinged, white, waxy little jewels. They can be very fragrant, with a spicy scent. Metallic blue fruits develop after the flowers, maturing to a deep black or blue-black. The fruits are persistent and add a new dimension of visual interest to the evergreen foliage.

Laurustinus viburnum retains its foliar beauty throughout the year, which makes it an ideal plant for use in mass or as a hedge or screen. It is hardy to zone 8 but will manage in protected areas of zone 7. It has the valuable attribute of tolerating shade, and even prefers some afternoon shade in an area with hot summers. Try planting it in a sheltered site where there is shade during the long summer days and relief from wind and cold in the winter. *Viburnum tinus* grows best in well-drained soil and may suffer from root rot in very wet areas. This viburnum is relatively drought-tolerant and should be long lived without undue site stress.

Cultivars of laurustinus viburnum include 'Variegata', with creamy yellow variegation, and 'Compactum' (Plate 47), which grows to only half the size of the species and sports dark red buds instead of the usual pink buds. Propagation is relatively simple. Summer softwood stem cuttings root readily under mist.

This useful and beautiful plant is chronically underutilized. The clean, glossy foliage of *Viburnum tinus* makes an attractive foil for the color of early spring flowers and bulbs, and its sweet blooms add their own enjoyable touch to the garden.

Weigela florida / Weigela   PLATE 48

This shrub conjures up images of grandmother's front yard: white picket fence draped with honeysuckle, wisteria clambering up the porch pillars, and flanking the walk in the yard, such old Victorian-era favorites as beautybush (*Kolkwitzia amabilis*), deutzias (*Deutzia* spp.), and old-fashioned weigela, *Weigela florida*. But before horticultural sophisticates bring an arsenical attitude to such an old-lace landscape, consider the reasons for the popularity of such shrubs, especially *Weigela florida* in its dozens of modern incarnations.

Weigela is a close botanical relative of forsythia (*Forsythia* spp.) and is native to Japan from whence it was imported during the Victorian era. Like forsythia, weigela is a large, somewhat loose and relaxed shrub that, if left to its own devices, matures into an open, arching mass. Also like forsythia, the oval foliage of weigela is basically unremarkable, with the exception of a few variegated forms, and it has little or no fall color. The small, dry tan fruit and light brown bark of weigela, while not unattractive, are also less than distinguished.

So why was this shrub such a favorite with American homeowners for so long? For two reasons: its pretty flowers and its reliability. An abundance of pink, rose, or white trumpet-shaped flowers appear unfailingly each year in spring (and often sporadically throughout summer and early fall), decorating the arching branches of this unflappable plant, one of the most persevering-in-the-face-of-all-odds landscape shrubs to come out of the great era of plant exploration in the 1800s.

Completely hardy to zone 4, *Weigela florida* grows to a large, rounded mass 5–10 feet (1.5–3.1 m) tall, with an equal or greater spread as the plant matures and branches arch toward the ground. Its informal habit is, for the most part, best left to itself to develop its own relaxed outline (although periodic pruning to renew flowering vigor may be helpful on older selections). Weigela will grow in almost any location. While it prefers moist, well-drained soils in full sun, it will thrive just about anywhere except in dense shade. It is essentially pest- and disease-free and is very tolerant of pollution and urban conditions, but it does need some room to spread. Weigela is very easy to propagate, either from seed, which can be sown fresh, or from stem cuttings taken in early summer and rooted under mist.

The ease of propagation has contributed to a great proliferation of cultivars of weigela. In addition to *Weigela florida*, a number of other species have been used in hybrid breeding programs for selections of weigela both in Europe and North America. As a result, the modern gardener has an array of named weigela of both species and hybrid origin to choose from, with a rainbow of flower and foliage color and an increased diversity of growth habit and cold-hardiness.

A mere sampling of these cultivars includes 'Avalanche', a large, vigorous form with bright white blooms tinged with pink; 'Boskoop Glory', with very large, salmon pink blooms; 'Bristol Ruby', an extra-hardy, relatively upright form with deep ruby-red flowers; 'Candida', with flowers that are pure white from bud to bloom, and spring-green foliage; and 'Conquerant', with extremely large, deep pink flowers and a broad-spreading habit. The flowers of 'Dart's Colordream', a well-behaved, midsize cultivar, emerge both cream and rose on the same plant. 'Dropmore Pink' is one of the most cold-hardy forms, with baby pink flowers. 'Eva Rathke' is a compact, slow-growing form that bears carmine flowers with contrasting, bright yellow anthers. 'Evita' also has bright red flowers, but its habit is low and dense, reaching only about 2 feet (0.6 m) in height. 'Foliis Purpureis' is a relatively dwarf form reaching 3–4 feet (0.9–1.2 m) in height, with purple-tinged foliage and pink flowers. 'Gracieux' bears salmon-rose flowers with yellow centers, displayed on upright branches. 'Java Red' has flowers of a very deep red and also has some red in the foliage. 'Looymansii Aurea' is a variegated form with surprising gold-yellow leaves with a thin, dark red edge and pink flowers. 'Minuet' is one of the most compact forms, reaching only 3 feet (0.9 m) in height with eye-catching flowers; the petals are deep rose-red with a purplish interior and a yellow throat. 'Mont Blanc', a vigorous grower with large blossoms, is generally reported to be the best white-flowered form. 'Pink Princess' is a loose, spreading selection with lilac-pink flowers; 'Polka' is a medium-sized, compact form with dark green foliage and light pink blooms set off with yellow centers. 'Red Prince' is an especially hardy form with dark red flowers that hold their color very well, on relatively upright branches, and is very likely to rebloom throughout the season. 'Rubidor' is a technicolor extravaganza, with bright, canary-gold foliage and magenta-red flowers that can be almost too much but never fail to attract attention. 'Rubigold' is similar to 'Rubidor', but the variegated foliage is lime-splashed gold. 'Rumba', 'Samba', and 'Tango' are all compact selections reaching about 3 feet (0.9 m) in height with reduced spread. All three bear red flowers with yellow centers and show varying degrees of purpling in the foliage, with 'Tango' showing the greatest and most persistent degree of purple in its leaves. 'Vanicek' (considered the same as 'Newport Red') offers purple-red flowers and bright green winter stems—an advantage, because weigela (like forsythia) is often rather a liability in the winter landscape. 'Variegata' has leaves edged with cream and deep rose flowers, while 'Variegata Nana' is a small form of 'Variegata', reaching 3 feet (0.9 m) in height.

With this incredible selection of weigelas, there is every reason to include at least one in any garden except the smallest. This reliable shrub is a rewarding addition to a mixed border, where the ebullient flowers can shine in spring, and its more modest character the rest of the year is not a detriment. The compact, variegated forms are excellent patio and container plants and are often good solutions for what to do with awkward corners. While all of the new selections are rightfully very tempting, there is also an old-fashioned appeal to a well-grown *Weigela florida* lounging in the middle of a lawn, spreading armloads of unabashed magenta-pink blooms.

Wisteria spp. / Wisteria   PLATES 49, 50

Wisteria has a reputation in the southern part of the United States for being rather voracious when it comes to growth. It is indeed an exceptionally vigorous climber but with some controlled pruning and training, its opulent beauty and heady fragrance can be a well-behaved addition to the garden.

The several species of wisteria found in the United States are similar in appearance and a bit tricky to distinguish, but they are all woody vines whose flowers are borne in showy, pendulous clusters each spring. All wisterias are climbing, twining vines with flowers of various shades of lavender-purple or white. The flowers emerge from early to late spring, before or just as the leaves emerge. They all have divided leaves which remain a glossy grass-green in the summer and may develop some yellow fall color. The flowers mature into beanlike seedpods covered with a downy fuzz, an intriguing addition to the late summer and early fall landscape. Wisteria can be readily propagated from softwood cuttings rooted under mist, but to speed development, many cultivars are propagated by grafting onto seedling rootstock.

There are four primary species of cultivated wisteria, one native to the southeastern United States and three from Asia. All of the wisterias discussed here are hardy to at least zone 5 and will perform well in a range of soil conditions from rich, well-drained loam to heavy clay. Wisterias need full sun and a light hand with the fertilizer for best flowering. To prune for control, cut the vine's vigorous stem growth back to three or four main buds. Root pruning can also be an effective restraint. Root prune by slicing into the soil to a depth of 8–10 inches (20.3–25 cm) with a sharp spade, in a circle or line 3–4 feet (0.9–1.2 m) out from the vine's crown (this distance will increase as the plant ages). The closer to the plant and more complete the circle, the more severely the growth will be inhibited.

Wisteria will happily climb up anything that comes within its reach with great abandon. The vines can be trained along arbors, trellises, and walls, and even up the sides of buildings. Wisteria vines can also be trained to a tree form by securing the plant to a fixed column or pole as it grows and continually pruning back new stem growth to a particular crown size. Eventually, after the trunk has matured, the support column is removed.

Probably the most common wisteria available is *Wisteria floribunda*, Japanese wisteria. It is somewhat more cold-hardy than other wisteria, being reliable to zone 4. Native to Japan, this species has received the most horticultural attention in terms of cultivar development, with a large number of cultivars available. Japanese wisteria can be distinguished from its cousins by two basic characteristics: The main stem will, in general, twine in a clockwise direction, and the flowers will open gradually along the cluster, from the base to the tip, as opposed to opening all at once. Most of the plants seen escaping up the trees of our woods and fields are Japanese wisteria.

The cultivars of Japanese wisteria are generally preferable to seedling plants, and they include a range of bloom color. 'Alba' is a white flowered form; 'Honey Bee Pink' is a very prolific, cherub pink form with good fragrance, usually a showstopper; 'Issai' has extremely long clusters of violet flowers, potentially reaching 24 inches (60 cm) in length; 'Ivory Tower' has prolific white flowers with an intense aroma; 'Kyushaku' has red-violet flowers; 'Lawrence' offers pale blue blooms with white centers; 'Longissima Alba' has long, clear white flower clusters with good fragrance; 'Rosea' is a pale rose-violet form with a luscious fragrance and especially long flower clusters; 'Royal Purple' is a deep purple with only a light fragrance; and 'Violaceae Plena' is a double-flowered, violet to violet-blue form with somewhat shortened clusters and a light fragrance.

The second most common wisteria is *Wisteria sinensis*, Chinese wisteria. This species is almost identical to Japanese wisteria, except that the flowers are more blue-violet than true violet, and all of them open simultaneously along the cluster. Also, the vine twines in a counterclockwise direction. Chinese wisteria may flower a bit later than Japanese wisteria. There are fewer cultivars. Among them are *W. sinensis* var. *alba*, a white-flowered form that comes true from seed; 'Black Dragon', a very dark purple, double-flowered form; and 'Plena', a lavender, double-flowered form.

Wisteria venusta or silky wisteria is the least common of the Asian forms. Silky wisteria is a demure, white-flowered form with smaller, more rounded flower clusters that are usually 4–6 inches (10.1–15.2 cm) long and nearly as wide. The flowers open all at once. The beanlike seedpods are covered with long, silky down.

Our native wisteria in the eastern part of the United States is *Wisteria frutescens*, American wisteria. Later flowering than the Asian wisteria, the American species is also seen in woods and fields but is not as competitive, so is less common in the wild. The flowers are pale lavender with a soft yellow center and are borne in smaller, more rounded clusters than those of the Asian wisterias. It is also less vigorous in the garden than Chinese or Japanese wisteria. There are white-flowered forms of American wisteria, one of which is 'Nivea'. Another native wisteria is *W. macrostachya*, native to the upper Midwest. It can be seen in the fields and woods from Arkansas to Illinois.

The wisterias offer a languid opulence of flower, form, and aroma that few climbing plants can bring to the spring landscape. There is something almost operatic about the sight of a wisteria draping an arbor with its massive clusters of bloom. Wisterias deserve a place where gardeners and visitors alike can be entranced by the floral coloratura of these horticultural divas.

Zenobia pulverulenta / Dusty zenobia   PLATE 51

Among the native plants well-adapted to many modern landscapes is dusty zenobia, *Zenobia pulverulenta*. Dusty zenobia can be found growing in the wild from North Carolina south into Florida. In spring, this entrancing shrub is hung with hundreds of white, bell-shaped flowers with fluted edges.

Dusty zenobia is an open shrub with gracefully arching branching that will reach 3–6 feet (0.9–1.8 m) in height with roughly equal spread. *Zenobia* is a relative of our edible blueberries, and its flowers are similar in shape to those of blueberry, but they are much larger and showier than blueberry blooms, and have a light, spicy scent. The flowers are borne all along the branches and dangle from thin stalks in the axils of the leaves in late spring. Dusty zenobia's foliage is also very attractive. The oval leaves, 2–3 inches (5–7.6 cm) long, are variably covered with a waxy bloom that lends them a lovely silver-blue cast. Leaves develop striking yellow-orange fall color with purple, reddish hues.

The shrub is hardy to zone 5. In mild areas, it is likely to be semievergreen, but at the cold end of its range, the leaves will drop for winter. Like its blue-fruited cousins, dusty zenobia prefers moist, acid, well-drained soils. These conditions will give the most rapid growth, but *Zenobia* will also perform well in less ideal soils if the severe extremes of heavy, wet clay or poor, dry sand are avoided. Full sun will give best flowering and bluest foliar color, but *Zenobia* will also tolerate light shade.

Dusty zenobia is easily propagated from seed. The seedlings show variability in floriferousness and degree of blue bloom on the foliage. The best blue-foliaged forms can be propagated by softwood cuttings rooted under mist, but success with rooting can be unpredictable. Some specialty nurseries in the Southeast that work with native plants offer excellent forms of dusty zenobia.

As spring fades rapidly into summer, the flowers of *Zenobia pulverulenta* are an exquisite reminder of the ephemeral nature of each season in the garden. No matter your preference for native or exotic, dusty zenobia is a special plant for American gardens— one that rings with delight when the season turns toward summer.

Summer

Aralia spp. / Devil's walking stick, Japanese angelica tree PLATES 52, 53

Both *Aralia spinosa*, devil's walking stick, and *Aralia elata*, Japanese angelica tree, make strong statements in the landscape. Their exotic summer foliage and coarse, thorny winter stems make them an invaluable addition to any gardener's palette of form and texture.

Devil's walking stick (Plate 52), sometimes called Hercules'-club, is a large, coarse-stemmed deciduous shrub, 10–25 feet (3.1–7.5 m) tall, with a few main unbranched stems that grow in a cluster, forming a small stand. Each thumb-thick stem is covered with large (and painful) spines, 0.5–2 inches (1.2–5 cm) long, which undoubtedly is the characteristic responsible for the plant's common names. The shrub has an overall oval to rounded outline, but the texture of individual stems is prominent. The appearance is quite fascinating—as if some prehistoric jungle had been miniaturized and set in the landscape.

Aralia spinosa is native to the south-central and southeastern United States and is hardy through zone 4. In winter, the bare, light gray stems create strong silhouettes. The foliage is large and doubly divided, somewhat like that of Kentucky coffee tree (*Gymnocladus dioica*). The many leaflets of each compound leaf are 3–4 inches (7.6–10.1 cm) long and appear as if they are the individual leaves. A single true compound leaf reaches 2–3 feet (0.6–0.9 m) in width. The emerald-green foliage is very tropical in appearance and softens the stems' bold character to some degree in the summer. Fall foliar color is often a medium yellow but is very variable and not of primary interest.

In midsummer, masses of tiny creamy flowers emerge from the foliage, held in erect, branched clusters that add yet another dimension of form and texture. The flowers ripen to purple-black fruits suspended from rose-colored stalks.

Japanese angelica tree, *Aralia elata*, is very similar to devil's walking stick but is taller, reaching up to 40 feet (12.4 m) with time, and is perhaps a bit more refined in summer texture. It is native to Japan and Korea and is somewhat more cold-hardy than the American *Aralia spinosa*, growing successfully as far north as zone 3.

Aralia elata offers some beautifully variegated cultivars, a trait unavailable in *A. spinosa*, but the variegated forms are extremely difficult to find in the United States, as they must be grafted in one of the most difficult procedures involved in propagation of any ornamental plants. An especially beautiful form is 'Variegata' (Plate 53), which displays foliage splashed with irregular borders of eggshell-white, very refreshing in the summer landscape. 'Aureo-variegata' has leaves with light gold borders.

Both *Aralia* species are well adapted to a range of landscapes. They will perform well in sun or partial shade. They do best in moist, well-drained soils but will tolerate heavy clays. They are relatively rapid growers as young plants, but growth slows when mature wood develops. *Aralia spinosa*, especially, suckers readily and can be an invasive spreader if not controlled by weeding up the suckers; however, this suckering tendency can be a good way to develop a planting fairly rapidly. Propagation is by seed (which requires two to three months of chilling pretreatment), or by digging the root suckers. Cuttings taken from the roots themselves will also successfully root and develop shoots.

Form and texture are often less-advertised attributes of the garden, but they are just as vital as color. These aralias offer a combination of form and texture that is outstanding in both summer and winter. At the North Carolina State University Arboretum, *Aralia* are growing in a number of sites, including the white garden, illustrating the fascinations of these plants in the landscape.

Buddleia davidii / Butterfly bush   PLATES 54, 55

When it is in full bloom, for weeks during midsummer, the gracefully arching branches of the shrub *Buddleia davidii* are alive with butterflies, delicately moving from flower to flower on the clustered, fragrant blooms.

Butterfly bush is sometimes called summer lilac because the flowers are reminiscent of lilac blooms, but butterfly bush does not resemble lilac (*Syringa* spp.) overall. *Buddleia* is a relatively open, arching plant, hardy to zone 5, which is grown as a herbaceous perennial in northern parts of America. In areas with mild winters, it develops into a woody shrub reaching 10 feet (3.1 m) or more in height with a 6–8 foot (1.8–2.4 m) spread. The blue-gray foliage is somewhat coarse-textured but makes an attractive foil for the lilac-like panicles of flowers that nod from the ends of the branches.

Buddleia is easily transplanted, and it is a tough survivor. It prefers well-drained, fertile soils but will also thrive in clay and sand as well. It needs full sun for best flowering and growth, and should be pruned back to about a foot (0.3 m) from the ground early each spring for more vigorous growth and better flowering. *Buddleia* is readily propagated from untreated seed or softwood cuttings under mist; cuttings should be moved out of the mist once they have rooted.

Butterfly bush is a lovely plant for a mixed shrub border or for a mass planting combining some of the many different cultivars. It flowers later than many shrubs, adding color and interest to the midsummer landscape.

Dozens of cultivars of *Buddleia* are widely available. They flower at slightly different times and come in a range of colors. 'Black Knight' displays intense, deep purple flowers with an orange eye against vigorous growth, 'Charming' blooms in pretty pink clusters. 'Pink Delight' is a form from the Netherlands with large, deep pink flower clusters. 'Royal Red' produces rich purple-red flowers; 'Empire Blue' has deep violet-blue blooms with orange eyes; and both 'White Profusion' and 'White Bouquet' offer panicles of pristine white to the gardener and the butterflies. There are several compact cultivars with lower, spreading growth that reach only about 5 feet (1.5 m) in height. These are great choices for smaller gardens, but they still need full sun. 'Nanho Alba' has narrow leaves and white flowers; 'Nanho Blue' has blue-lavender flowers; and 'Nanho Purple' has purple flowers and gray-green foliage.

There are also a few interesting orange-flowered forms of Buddleia. *Buddleia globosa* (Plate 54) is a light-orange flowered species with small, globular flower clusters. The flowers are a surprising pastel orange (Plate 55), but the clusters are small and far less showy than other forms of the more commonly grown *B. davidii*. *Buddleia globosa* is cold-hardy only through zone 7. Hybrids of *B. globosa* with *B. davidii*, include the cultivars 'Golden Glow' and 'Moonlight', which produce creamy yellow to pale orange flowers, and 'Sun Gold', with bright yellow-orange flowers.

No matter what color the blooms are, all butterfly bushes attract many different kinds of butterflies. There's nothing more magical than a *Buddleia* in full bloom, its pastel blossoms decorated with a fluttering of jewel-toned butterflies, slowly opening and closing their wings in the summer sun.

Campis radicans, *C. grandiflora* / Trumpetvine   PLATE 56

That ugly chain-link fence around the swimming pool, the neighbor's severe stockade perimeter, that all-too-new trellis by the patio—all are vexing landscape problems with an easy solution called trumpetvine. *Campsis radicans*, trumpetvine or trumpet creeper, is a very rapidly growing vine that will quickly cover fences, walls, trellises, poles and anything (or anyone!) that stands in one place just a little too long. The vine is native to much of the eastern half of the United States and has naturalized as far north as the Great Lakes region along the Canadian border. In the wild, you will see it growing along fencerows, on telephone poles, and at the edges of woods and old fields.

Trumpetvine has glossy green, divided foliage, which turns a light yellow in the fall. The lustrous foliage is a perfect companion to the vine's rich orange flowers. Clusters of the narrow trumpet-shaped blooms are produced from summer through autumn, peaking in early summer. Not only are the mature flowers dramatic, but the unopened, rounded flower buds are also a beautiful, velvety orange, so that the entire flower cluster is very pleasing to the eye. As a bonus, the flowers are especially attractive to hummingbirds. The fruits of trumpetvine are of interest late in the season. The long brown pods dangle from the vine as the foliage assumes its modest fall color, adding an extra dimension to trumpetvine's appeal.

Campsis radicans is incredibly stress-tolerant and thrives in areas where other plants will not survive. Michael Dirr has gone so far as to remark in his *Manual of Woody Landscape Plants*: "If you cannot grow this, give up gardening." Trumpetvine has no major pest or disease problems and will grow in almost any soil or site. Full sun is required for prolific flowering, but the vine will also grow well and flower sporadically in part shade. It is cold-hardy through zone 4.

Because of its tolerance of harsh environments and its rapid growth rate, the native *Campsis radicans* can be extremely invasive. Judicious pruning every spring will control its rampant growth, if necessary. The vigorous vine is effective camouflage for many types of landscape surfaces. Few plants will cover an area as rapidly as trumpetvine, a quality which can be a great advantage in a new landscape site, or a great bane if accidentally introduced into a small garden. Make sure that supports are exceptionally sturdy in order to support the weight of trumpetvine's vigorous growth. When pruning, be careful to minimize contact with cut stems and foliage. Some people can develop a poison-ivy like rash after contact, which gives trumpetvine its other common name of "cow itch."

If you want the beauty of *Campsis radicans* without the potential for bionic invasiveness, an Asian species, *Campsis grandiflora*, provides the perfect alternative—with the added plus of flowers two to three times the size of the blooms of *C. radicans*. Chinese trumpetvine, *C. grandiflora*, is a decidedly slower growing species with foliage nearly identical to that of *C. radicans*. The flowers are large, brilliant tangerine trumpets with an especially broad and showy corolla. They cover the plants so dramatically that the incredible mass of color literally stops people in their tracks. The vine blooms heavily in late spring with sporadic rebloom throughout the summer. It is important to note that *C. grandiflora* is significantly less cold-hardy than *C. radicans*, being reliable through zone 7. Chinese trumpetvine is becoming more widely available through the joint introduction program of the North Carolina State University Arboretum with the North Carolina Association of Nurserymen.

There is a named hybrid of *Campsis grandiflora* and *C. radicans*, *C. × tagliabuana* 'Madame Galen' (Plate 56), with blooms intermediate in size and color between the small, darker flowers of native *C. radicans* and the larger, brighter flowers of *C. grandiflora*. 'Madame Galen' is also more cold-hardy than *C. grandiflora*, being reliable to zone 5.

Also available are some selected cultivars of the native *Campsis radicans*, which differ from the species mainly in terms of flower color. 'Flava' has yellow flowers; 'Praecox' produces rich red blooms.

All *Campsis* can be easily propagated from root cuttings, or from softwood cuttings taken in early summer and rooted under mist.

Whichever selection you choose, take advantage of the assertive nature of this robust plant to cover your landscape problems, from top to bottom, with a blaze of orange in summer. But before you plant, make sure to consider your space and your willingness to keep trumpetvine in check with regular pruning.

Carpinus spp. / Hornbeams   PLATES 57, 58

Hornbeam trees add year-round beauty and grace to the landscape, but it's during the dog days of summer that their refreshing green canopy is truly appreciated. A number of *Carpinus* species are excellent trees for the landscape. These deciduous relatives of birches (*Betula* spp.) reach from 20 feet (6.2 m) tall to as much as 60 feet (18.6 m), and spread to 30 feet (9 m). Slender branches arch up and away from their distinctive smooth-barked trunks. The silvery or blue-gray trunks are a good foil for the bright green leaves. The foliage is similar in appearance to that of birches, but the leaves are generally smaller and a bit more refined, often with better quality throughout the growing season.

The intriguing fruits of hornbeams look like hundreds of tiny papery whirligigs, strung together in dozens and dangled from the branches. These fruits are a good tool for identifying *Carpinus*, particularly when separating it from its lookalike cousin *Ostrya virginica*, hophornbeam. Hornbeam's fruit resembles a cluster of many single-winged maple samaras strung together on a thread. Unlike hornbeam, hophornbeam's fruit clusters resemble the fruit of the hops plant, a series of overlapping whorls of small, dry sacs that from a distance are reminiscent of a length of heavy lanyard.

Carpinus caroliniana, the American hornbeam, is the only North American native of this genus. It is often called musclewood, because the sinewy trunks bear an uncanny resemblance to the limbs of health-club devotees. *Carpinus caroliniana* is also called ironwood because of its very hard wood, traditionally used for mallets, golf clubs, and tool handles.

Musclewood is a small understory tree, generally 10–30 feet (3.1–9 m) tall. It is found in moist places throughout the eastern half of the United States and north into Canada around the Great Lakes and east along the St. Lawrence river basin, and it is no less lovely for its common occurrence. These trees can have spectacular fall color in shades of canary-yellow and scarlet, but this character varies throughout the seedling population, so purchase plants in the fall to preview their fall color.

The American hornbeam grows very well in shade and so is a good choice for creating scale in gardens under towering older trees. It also performs admirably in full sun, as long as there is adequate moisture. It will tolerate flooding and many other urban landscape conditions, even those near parking lots and streets, although other *Carpinus* are probably better choices for those very stressful locations. It is best to transplant the tree balled-and-burlapped because it does not tolerate transplanting well. American hornbeam is cold-hardy to zone 3, and may be somewhat more heat-tolerant than other *Carpinus*.

Of the hornbeams, *Carpinus caroliniana* is the most likely to suffer disease and insect problems on the foliage, but these are usually not severe. The tree is somewhat difficult to propagate, both from seed and vegetatively. It is generally propagated from seed collected while green and stored in a moist medium at 40°F (4°C) for four months.

An uncommon cultivar called 'Pyramidalis' has an inverted pyramidal habit of variable character.

Among the nonnative hornbeams, *Carpinus betulus*, the European hornbeam, is the most widely grown (although no hornbeams are especially common in the trade). It is very similar in appearance to American hornbeam but is somewhat larger, reaching 50–70 feet (15.2–21.7 m) with age, with larger leaves and a more upright, pyramidal habit. The foliage of this beautiful tree remains clean and clear green all summer. In winter, it displays an elegant tracery of branches above its smooth bark.

European hornbeam is not a rapid grower, making it a good choice for size-restricted areas, especially if some of the upright cultivars are used. It is cold-hardy to zone 4. It does best in well-drained sites in full sun but also performs well in heavy clay soils.

This tree is a beautiful choice for street boxes, parking lot planters, and low-walled condominium patios. It remains dignified twelve months of the year, even under city conditions.

A number of cultivars of European hornbeam are superior to seedlings for landscape use. Two cut-leaf forms are 'Asplenifolia', with a lacy texture created by deeply lobed foliage (similar to that of cutleaf beech, *Fagus sylvatica* 'Laciniata'), and 'Incisa', with shorter foliage that is more coarsely lobed. 'Columnaris' and 'Fastigiata' are upright, narrowly columnar forms that are often confused in the trade; both are fine-quality selections with nearly identical landscape character, perfect for street trees, hedges, patio specimens, or anywhere a compact tree is needed. 'Globosa' is a very rounded form—almost a large shrub—that makes a good hedge or screening plant. 'Horizontalis' has a flattened crown, while 'Pendula' (Plates 57, 58) has a weeping habit. The new foliage of 'Purpurea' emerges burgundy but quickly changes to green, especially in areas with hot summers. The leaves of 'Quercifolia' are reminiscent of oak foliage. Those of 'Variegata' are splotched with light yellow, and 'Albo-variegata' leaves are marked with white. In the west half of the North Carolina State University Arboretum, a number of different *Carpinus* display their impeccable character where any gardener can experience the green refreshment of hornbeam in full summer leaf.

Propagation of European hornbeam is relatively difficult both from seed and from cuttings. Named selections are usually grafted onto seedling understock, and researchers have worked to improve vegetative propagation techniques. In the past, these difficulties have contributed to the relative scarcity of hornbeams in commercial production, but as more growers work with grafting and specialty crops, hornbeams become more widely available.

The demands of summer's heat and changeable moisture conditions make it difficult for trees to keep their fresh green character throughout the season. Hornbeams meet the challenge of the dog days with sprightly green leafy canopies and cool silver trunks.

Caryopteris × clandonensis / Blue-mist shrub   PLATE 59

Nothing is more soothing than relaxing in the garden with a tall glass of minted iced tea and a view of fresh, green leaves with cool, blue flowers scattered among the foliage. Blue-mist shrub, *Caryopteris × clandonensis*, gives us just such an oasis in our gardens.

Caryopteris is a genus of low-growing deciduous shrubs that flower in summer with small, frothy clusters of blooms in shades of blue and lavender. The most widely grown *Caryopteris* is blue-mist shrub, actually a hybrid of two species.

Blue-mist shrub is a hybrid of *C. incana* and *C. mongolica* that was originally raised in the garden of Arthur Simmonds in England. The hybrid is a low shrub, about 2 feet by 2 feet (0.6 m by 0.6 m), with lavender-blue flowers in small clusters scattered throughout the foliage. Of its two parent plants, *C. incana* is more widely available in the trade. It is a small shrub with fine, narrow leaves, covered with a light gray pubescence and bearing lavender flowers. *Caryopteris mongolica* is a dwarf, scraggly plant with blue flowers.

The most common cultivar of the hybrid *Caryopteris × clandonensis* is 'Blue Mist', a low, mounded, deciduous shrub with light-textured, narrow foliage, reaching 2–3 feet (0.6–0.9 m) in height with an equal spread. Feathery clusters of delicately scented, powder-blue flowers peek out from the leaves.

An older cultivar with medium blue flowers is named 'Arthur Simmonds' in honor of the hybrid's discoverer. A number of other gardenworthy cultivars of *Caryopteris × clandonensis* are also available. 'Dark Knight' has extremely dark lavender flowers—almost deep purple. 'Longwood Blue' is a larger, especially floriferous cultivar with soft blue flowers, from Longwood Gardens in Kennett Square, Pennsylvania. It reaches 4 feet by 4 feet (1.2 m by 1.2 m).

While blue-mist shrub is a cooling influence in the landscape, it is actually a heat-loving plant. As a result, it does very well in areas with long, hot summers, but it also tolerates some winter cold, being hardy through zone 6. *Caryopteris × clandonensis* prefers full sun and well-drained soil. It can be easily propagated from softwood cuttings rooted under mist. Because the flowers are borne on new growth, this shrub can be treated as an herbaceous perennial and pruned back each winter to stimulate plenty of new growth in the spring.

Caryopteris incana, the predominant parent of *C. × clandonensis*, is a great garden plant in its own right. Common bluebeard is somewhat more loose and open than blue-mist shrub and reaches 3–5 feet (0.9–1.5 m). It is also slightly less cold-hardy (to zone 7) than blue-mist shrub, it but can be grown as an herbaceous perennial or die-back shrub farther north where the roots are cold-hardy but the top dies back to the ground each winter. It is as easily propagated as other *Caryopteris*. A brilliantly colored cultivar of *C. incana*, 'Worcester Gold' (Plate 59), carries bright canary-gold foliage that is a spectacular combination with its lavender-blue flowers.

The light foliage and delicate blue flowers of any of the *Caryopteris* make these shrubs a perfect choice to cool and refresh a hot summer landscape. The fine texture of foliage and flowers make blue-mist shrub easy to use in many locations, from patio container to perennial border to naturalized mass. One thing is certain: Every summer garden will say "Ahhh" to the cool blue-mist shrub.

Cedrus spp. / True cedars   PLATE 60

At hot, humid times of the year, nothing is more cooling than a glimpse of blue-green foliage. There are many garden plants with bluish foliage, some of which hold their blue color reliably throughout the season and others which fade in the heat. _Cedrus deodara_, deodar cedar, and _Cedrus atlantica_ 'Glauca', blue Atlas cedar, are two beautiful conifers that add cool shades to hot summer landscapes and graceful form to the garden year-round.

Cedars are large, conical to spreading conifers with fine-textured needles, which lend them a softer appearance in the landscape than other conifers such as pines and spruces. Native to various parts of Asia and the Middle East, they will reach 40–80 feet (12.4–24.8 m) in height and will spread with age. Cedars are excellent conifers for many parts of the United States, especially the Southeast, where many other conifers will not survive the hot, wet summers. True cedars make excellent large specimen trees. They are tolerant of a wide range of landscape conditions, except for extremely wet sites.

Three species of true cedar (not to be confused with eastern red cedar, _Juniperus virginiana_ or certain species of _Thuja_, _Chamaecyparis_, and other conifers also commonly known as cedars) are available in the landscape and nursery trade: _Cedrus deodara_, _C. atlantica_, and _C. libani_.

Cedrus deodara is often used as a large landscape tree. Young trees are conical but grow to be flat-topped and spreading with age. Frequently, the main leader dies out with age, contributing to their spreading habit. Deodar cedar is the most graceful of the three cedars treated here, with pendulous branching and blue-green needles, especially in youth. In midsummer, light blue-green, upright cones, swelling as they mature, decorate these trees like Christmas ornaments. This species is hardy through zone 7, but 'Shalimar' is an especially cold-hardy, full-size cultivar introduced by the Arnold Arboretum that performs well in the Boston, Massachusetts, area and through zone 6. At the other end of the spectrum, 'Nana' is a small, mounded dwarf with silvery gray-green foliage—an excellent choice for southern rock gardens but not suited to cold-winter gardens because it is reliably hardy only to zone 7, and 'Pendula' (Plate 60) is a beautifully weeping form.

Cedrus atlantica, Atlas cedar, has thicker, stouter needles and is somewhat more cold-hardy than _C. deodara_, being reliable through zone 6. Although Atlas cedar is most familiar in its blue-needled forms, the needles of the species are darker green than those of Deodar cedar. As a young tree, the habit is open and stiff, but the tree becomes flat-topped with age, assuming picturesque horizontal branching. Two very beautiful cultivars of Atlas cedar are widely used. _Cedrus atlantica_ 'Glauca', the blue Atlas cedar, is strikingly blue (although there can be a certain amount of variation in color from source to source), because the needles are covered with an extra-thick layer of wax. _Cedrus atlantica_ 'Glauca Pendula' is an extremely weeping form of the blue 'Glauca'. The branches of this cultivar will cascade from an arbor like a waterfall. There are many beautiful trained specimens across the country, in which the main leader of the tree has been trained horizontally across a trellis or entryway or along an arbor in such a way that a waterfall of secondary branches rains down from the supporting structure. At the North Carolina State University Arboretum, a lovely plant of this cultivar drapes the entrance of the visitors' center.

A third cedar available in the trade is _Cedrus libani_, cedar-of-Lebanon, which is indeed the famous tree of Biblical times. The foliage is often darker green than that of its two cousins, to the point of looking almost black-green in bright sun. Once seen, the incomparably beautiful cedar-of-Lebanon is rarely forgotten. It grows from an openly pyramidal tree when young to a flat-topped, horizontally spreading tree. At the Orlando E. White Arboretum (Blandy Farm) in Boyce, Virginia, a spectacular allée of *C. libani* lines one of the drives. Unlike the other *Cedrus*, cedar-of-Lebanon can be more of a challenge to grow, doing best in a pollution-free, dry, sunny location with well-drained, loamy soil. Cultivars include 'Argentea', a silver-blue form, and 'Pendula', a weeping form that is often grafted to produce a small weeping specimen. *Cedrus libani* var. *brevifolia* (Cyprus cedar) is more diminutive than the species in all its characters. *Cedrus libani* var. *stenocoma* is the most cold-hardy *Cedrus* (to zone 5). 'Sargentii' is a dwarf form of *C. libani* with pendulous habit (but not quite as severely weeping as 'Pendula').

All cedars can be propagated from cuttings taken in fall, treated with rooting promoter, and rooted under mist, but they are slow to root and can give an unpredictable response. As a result, many cultivars are grafted in commercial production to speed the process and provide uniform, vigorous roots. *Cedrus libani* is perhaps the most difficult to root but can also be propagated from seed, as can the other cedars. Seed exhibits little or no dormancy, but a cold treatment for two weeks at 40°F (4°C) improves germination.

Cedars make beautiful large structural and sculptural trees for landscapes that can afford them a little space. Their soft blues and grays create a halo of cool like a distant mountain range against city horizons. Consider planting the graceful cedar to bring fresh majesty to the garden year-round.

Cercidiphyllum japonicum / Katsura tree   PLATE 61

The leaves of katsura tree are its main delight, hanging languidly from the branches and fluttering as easily as poplar leaves in the slightest breeze. The beautiful foliage emerges a red-wine color in the spring and rapidly changes to a fresh blue-green for the late spring and summer. The fan-shaped leaves of *Cercidiphyllum japonicum* are similar in outline to the leaves of redbuds (hence, the botanical name *Cercidiphyllum*, after *Cercis*, the botanical name for redbud). In autumn, the 3-inch (7.6-cm) leaves turn a remarkable clear yellow, sometimes tinged with pastel orange. A unique characteristic of katsura tree foliage is the strong, spicy fragrance that is released just before the leaves drop in the fall.

The habit of this large, deciduous tree is as pleasing as the foliage. Softly arching branching creates a pyramidal habit in youth, which generally matures to a more widely spreading crown with age. Katsura tree ultimately reaches 40–80 feet (12.4–24.8 m) in height but in cultivation is usually seen in the range of 20–40 feet (6.2–12.4 m), with a canopy spread of 20 feet (6.2 m). The smooth tan bark of young trees is a good foil for the foliage during the growing season and is appealing in its own right in winter. As trees age, the bark develops a more furrowed and shaggy character with some peeling of the bark. The inconspicuous greenish flowers of katsura tree emerge in early spring before the leaves. Male and female flowers are borne on separate plants, and so trees of each are needed for fertile seed development. Successfully fertilized flowers mature into interesting winged seeds in dry pods.

Katsura tree can be readily grown in a range of landscape conditions. It is winter-hardy to zone 4 and tolerant of most soils as long as it receives adequate moisture and full sun. It is sometimes difficult to move and should be balled-and-burlapped for transplanting. Regular watering during the first year is vital. The tree has no serious pest or disease problems. Katsura tree may be readily propagated from seed, which requires no special treatment; from cuttings taken in summer from very young trees and rooted under mist; or by grafting, which is the method used to propagate cultivars.

A second katsura tree, _Cercidiphyllum magnificum_, has larger, more rounded leaves than _C. japonicum_. Some botanists classify the two plants as separate species; others call the second tree a botanical variety of the first: _Cercidiphyllum japonicum_ var. _magnificum_. Both _Cercidiphyllum japonicum_ and _C. magnificum_ are Asian in origin.

There are a few cultivars of katsura tree, but they are not as easily found as the seedling trees. _Cercidiphyllum japonicum_ 'Pendula' (Plate 61) is a weeping form with branching cascades of foliage flowing right to the ground, while _Cercidiphyllum magnificum_ 'Pendulum' is a weeping form of the larger-leaved species that may get larger than _C. japonicum_ 'Pendula'. Both weeping forms are exceptionally beautiful in the garden and make smaller specimen trees than the species, reaching 15–25 feet (4.5–7.5 m) instead of 40–80 feet (12.4–24.8 m).

The handsome katsura tree has a rare combination of generous proportions and distinctive elegance in all four seasons of the year. Its adaptability to a wide range of climates and soils makes it the perfect choice to drape a garden with entrancing falls of blue-green foliage and graceful branches for a year-round curtain of woody beauty. The weeping katsuras are especially graceful additions to small gardens as unique but reliable specimens.

Clerodendron trichotomum / Harlequin glory-bower   PLATES 62, 63

Harlequin glory-bower is a loose, somewhat unkempt shrub that redeems itself with its lovely flowers and brightly colored fruit. It begins blooming in midsummer when few other flowering shrubs offer color and continues flowering until first frost. The strongly fragrant flowers are a soft white with a prominent red calyx at the base of each flower. The red calyx persists as the fruits develop into small, metallic blue berries, making a very colorful combination.

In pre-Linnaean Europe, before plants were assigned scientific names, two different _Clerodendron_ were known, probably _C. trichotomum_ and _C. bungei_. One of the two was believed to be the "tree of good fortune," and the other was believed to be the "tree of bad fortune." The reputation lives on in the botanical name of the genus, _Clerodendron_, which originates from the Greek _kleros_, "chance" or "destiny," and _dendron_, "tree." Unfortunately for modern gardeners down on their luck, we have lost the knowledge of which was which.

The genus _Clerodendron_ contains many species scattered primarily throughout the tropical areas of the world. _Clerodendron trichotomum_, harlequin glory-bower, is one of the relatively hardy species. A native of China and Japan, it is a loose, open, somewhat unkempt multistemmed shrub, which eventually reaches about 5–10 feet (1.5–3.1 m) tall and spreads to the same width. The overall shape is rounded but the open habit makes the shape a bit irregular. The plant can be trained to a small, multistemmed tree form by judicious pruning and diligent removal of suckers. Large, dark green, relatively coarse foliage is scattered to the ground along slightly fuzzy branches. The flowers are long and tubular, opening into five relaxed petals surrounding exceptionally long stamens. The flowers have an airy grace and are large enough to make an eye-catching show. The two-tone fruits of harlequin glory-bower are, to my eye, even more interesting than the flowers, although they require a closer look to fully appreciate them (Plate 63). The leaves quietly drop in the autumn with no show of fall color and may still be green when they fall.

Harlequin glory-bower is hardy through zone 6 and can be grown in zone 5 as an herbaceous perennial or die-back shrub. Cut stems to within about 10 inches (25 cm) of the ground after leaf-fall, and mulch. The shrub prefers moist, well-drained soils but also grows well in clay soils. Plant shape, flowering, and fruiting are best in full sun. The shrub will grow and bloom in partial shade, too, although the already loose habit will be even more open there. Propagation is easy from seed or cuttings taken from partially hardened wood in early summer and rooted under mist.

Clerodendron trichotomum var. *fargesii* is a botanical variety, native to China and Taiwan, that is similar to the species but has smooth stems and lacks the red color in the calyx. This variety has such strong cloying, musky fragrance that it can be offensive to sensitive noses.

Clerodendron bungei, the French hydrangea or red Mexican hydrangea, is taller and more upright than *C. trichotomum*. It has large, heart-shaped leaves and rounded clusters of purple-pink flowers similar to those of *C. trichotomum*, except that the individual flowers are smaller and are borne in large heads. *Clerodendron bungei* is not as cold-hardy as *C. trichotomum*, being reliable through zone 7.

An informal flowering shrub, harlequin glory-bower is well suited to use in a mixed border or as a grouped planting of several shrubs behind other planting beds. Because the foliage is not especially dramatic, it blends well with other plants, but the shrub adds its own beautiful flowers and unusual blue fruit to any planting at a time of year when few woody plants offer such a striking show. *Clerodendron trichotomum* also makes a good choice to include in a planting of blue conifers, where its shiny blue fruits and bold foliage add a change of interest and texture to the garden.

Clethra spp. / Clethras   PLATES 64, 65

Languid summer evenings are a time when the garden becomes a luxurious retreat. The heat of the day slowly fades with the light and the outlines of shrubs and trees take on a restful silhouette, punctuated here and there by the intensified glow of light-colored blooms and foliage. The humid air drapes lush fragrances around gardeners who are out and about at the end of the day—an especially delicious reward for an evening stroll (or an extra enticement for that overdue sunset weeding session). Among the most delightful of the many plants that contribute to the sensuality of summer's landscape are the species and cultivars of the genus *Clethra*.

Clethras are small shrubs to large trees with beautiful, dark green foliage and the unique attraction of showy flowering in mid- to late summer when few other shrubs or trees are in bloom. Some are very fragrant, others have exquisite bark character and all are excellent, relatively trouble-free garden plants. They are tremendously adaptable plants thriving in shade or full sun (which gives best flowering), with almost no pest or disease problems given adequate moisture. All are completely hardy through at least zone 5, and one, *C. alnifolia*, is surprisingly salt tolerant—making it an excellent plant for coastal gardens. Propagation of *Clethra* is quite reliable from either summer cuttings or ripe, dry seed.

Perhaps the best known of the *Clethra* species is the deciduous *Clethra alnifolia*, sweet pepperbush. This aptly named native shrub bears small fruits reminiscent of the dried black peppercorns that stock peppermills, and its flowers have a sweet and spicy fragrance. Sweet pepperbush can be found growing in wet areas up and down the eastern part of the United States from Maine to Florida, and west to parts of Texas. It is the most cold-hardy of the *Clethra*s discussed here (to zone 3). It is a low to medium-sized, rounded shrub reaching 3–8 feet (0.9–2.4 m) in height with distinctly upright branching. Its branches are covered with glossy, dark green foliage of neat, oval outline. In nature, the plants tend to sucker and develop into quite large, well-defined and attractive colonies. In midsummer, the ends of the branches are covered in upright clusters of small white blooms. More wonderful is the fragrance of these blooms, which can be a great pleasure even from a distance. In the fall, the leaves turn burnished gold and the flowers mature into the black, peppercorn-like capsules that are attractive in their own right.

Several notable cultivars of sweet pepperbush make excellent summer-blooming plants for many uses, from mixed border to specimen to naturalizing. 'Creel's Calico' is among the most spectacular, with large splashes and specklings of creamy white on the foliage; because the cultivar is quite new, it may be difficult to find in the trade). 'Hummingbird' (Plate 64) is an exciting, relatively new form with compact, almost dwarf habit, reaching 3 feet (0.9 m), and very prolific flowering. 'Paniculata' is an especially vigorous selection with large flower spikes. 'Pink Spires' (Plate 65) is a pink-flowered form whose buds emerge soft pink and lighten to pale pink as the flowers open, and 'Rosea' develops pink flower buds that open into pinkish white blooms.

In addition to the well-known *Clethra alnifolia*, several other species offer equally excellent garden character. *Clethra acuminata*, cinnamon clethra, is another eastern North American native found primarily in drier, montane areas only in the Southeast. It is slightly larger and more open than *C. alnifolia*. Its pendulous flower clusters are not quite as showy en masse as those of *C. alnifolia*, but cinnamon clethra has the additional attribute of developing warm reddish brown, peeling bark on older plants. *Clethra barbinervis*, Japanese clethra, is a large shrub or small tree from Japan that gets significantly taller than other forms, reaching upwards of 15–18 feet (4.5–5.6 m) with age. It has larger, more leathery leaves than its North American relatives, and its flower clusters are longer and semipendulous. The foliage develops vivid fall color in rusty red and gold shades. The bark of Japanese clethra has been compared with that of the magnificent *Stewartia* species, with its polished, warm tones and peeling character. *Clethra tomentosa*, wooly summersweet, appears nearly identical to *C. alnifolia* (and is sometimes classified as a botanical variety of *C. alnifolia*), but it has woolly stems, covered with a fine down, and a silvery cast to the foliage. It is found in the wild only in the southeastern United States. *Clethra macrophylla* and *C. pringlei* are rare species from Mexico, usually encountered only in arboreta and botanic gardens, with hardiness still in the testing stage. They show somewhat larger leaves than other forms and bear beautiful, lacy flowers.

Clethras can make important elements in a number of different kinds of plantings. They are wonderful as a fragrant mass or as delightful elements in a bog or stream garden, and they are excellent in a mixed border. The different forms of *Clethra* are all fine garden plants that bring sparkle and interest to the height of summer, with fresh flowers and spicy scent. Their foliage and bark add dimension to the garden in unexpected ways. In your next midsummer evening walk through your garden, imagine wandering the paths to find the dreamy flowers and fragrance of *Clethra* around each shadowed turn.

Cotinus coggygria / Smoketree PLATE 66

As the green of summer softens the outlines of trees and shrubs, the beautiful, old-fashioned smoketree, *Cotinus coggygria*, brings its own special grace to the landscape.

This large shrub or small multistemmed tree carries a cloud of airy flower clusters that float like pink gossamer above the landscape.

Native to southern Europe, central China, and the Himalayas, *Cotinus coggygria* has been in cultivation since the 1600s. It was a favorite of Victorian-era gardeners. Smoketree can grow to 15 feet (4.5 m) in height with a 10-foot (3.1-m) spread. It has an open, spreading habit and will achieve a pleasing, upright oval shape if left unpruned. Older plants can be limbed up to create multistemmed small trees.

The marvelous effect of "smoke" is due to the thousands of tiny pinkish hairs attached to the developing fruit, which are borne in delicate, branched clusters at the ends of the branches. Since the hairs are so tiny, each branch appears to end in a mist.

Smoketree flowers on the new, current season's growth. It can also be cut to the ground each winter, a technique that keeps the plant small enough to be used in restricted spaces or even in a perennial border. With this technique, growth can be amazingly vigorous, producing shoots that reach 3–5 feet (0.9–1.5 m) long.

Smoketree is reliably hardy to zone 5, and can be grown as a die-back shrub into zone 4. It makes a beautiful backdrop in a mixed shrub border and can be incomparable in bloom when massed in front of dark evergreens. The leaves of the species are a medium green. Some cultivars boast purple foliage, and some can develop good fall color. Both flowers and foliage are dramatic when cut and used for indoor floral arrangements.

Smoketree has no serious disease or pest problems, although it can rarely be affected by some rusts, leaf spot, and San Jose scale. It prefers well drained, full-sun sites but performs well in many soils, from light sands to heavy red clays. In water-conscious gardens, its excellent drought tolerance is a distinct advantage.

Propagation, both vegetative and from seed, can be successful but is tricky. Softwood cuttings are treated with rooting promoter and rooted under mist. Rooted cuttings are overwintered in the rooting medium and not potted up until the following spring. Propagation from seed is also not simple. Seed is soaked in acid for approximately 1/2 to 1 hour, succeeded by three months of chilling at 40°F (4°C). The seed is sown in fall and germinates in spring.

A number of cultivars of smoketree have special characters well worth seeking out. Several cultivars have deep burgundy-purple foliage with wine-colored flower panicles that are darker and richer in color than the type. The purple cultivars make a much bolder statement in the landscape than the light pink varieties. Their deep velvet foliage and burgundy blush of flowers is beautiful in combination with silver or blue-gray foliage plants, such as some willows or blue conifers. One of the best of the purple cultivars is 'Royal Purple', which can be so dark that the foliage appears almost black. 'Nordine Red', a spectacular wine-red cultivar, is particularly cold-hardy and often holds its color intensity better than others in areas with hot summers. 'Nordine Red' also has good orange and yellow fall color. Of the pink cultivars, 'Daydream' (Plate 66) is one of the most interesting, with uniquely dense inflorescences, relatively compact growth, and attractive green foliage. 'Purple Splendor', 'Velvet Cloak', and 'Pink Champagne' are also good cultivars.

In addition to these cultivars, there is another species of *Cotinus*, native to parts of the southern United States. *Cotinus obovatus*, American smoketree, is similar to the common smoketree but grows much larger, to 30 feet (9 m), and has spectacular fall color, ranging from brilliant oranges and purples to vibrant golds and carmines. It has beautiful gray bark and light pink flower panicles that are almost as showy as those of the common smoketree. It is hardy to zone 4. A hybrid cultivar, called 'Grace', is the result of a cross between *Cotinus coggygria* and *Cotinus obovatus*. Developed by Peter Dummer of Hillier Nurseries in England, 'Grace' has exceptionally long and massive flower panicles, which can be as long as 14 inches (45.5 cm) and as wide as 11 inches (27.8 cm), and it offers bright fall color. Unfortunately, 'Grace' is essentially unavailable at garden centers, although it can be ordered from a few specialty nurseries in American markets.

Smoketree is an old favorite with gardeners of an earlier generation, and it is rightfully regaining its place in modern gardens. As you drive along past older gardens and parks, you may see smoketree's delightful halo mingling with the summer afternoon's haze. Stop a minute to savor its graceful beauty and consider planting smoketree to add its gossamer blush to your own summer garden.

Cunninghamia lanceolata / China fir   PLATE 67

Many of the "old-fashioned" plants considered specialty, novelty, or choice plants for the garden today were imported from Asia during the plant-collecting expeditions of the 1800s. A number of the nursery owners, gardeners, and landscape designers of that era were especially interested in amassing collections of rare and unusual plants. Their clients were frequently royalty or the wealthy residents of large estates concerned with making a strong impact on their landscapes by using new, visually arresting or bizarre plants to create a "plant zoo" within the framework of the landscape design. One example of the legacy left from this era is China fir, *Cunninghamia lanceolata*. Named in honor of John Cunningham, who discovered the tree in China, China fir is not a true fir at all but is actually a Chinese relative of both our native bald cypress (*Taxodium distichum*) and Japanese cedar (*Cryptomeria japonica*), all of which are in the same botanical family, Taxodiaceae.

China fir is a large evergreen, coniferous tree with conspicuous flat needles that are as sharply pointed as a lance (hence, *lanceolata*). The shape of the needles is reminiscent of the shape of the needles of many of the true firs (*Abies* spp.). China fir grows with open, somewhat pendulous branching in a pyramidal habit when young, but it can become somewhat ungainly when older. Plant quality and habit and texture of the foliage varies tremendously from seedling tree to seedling tree, but China fir is always an exotic, unusual-looking plant.

Individuals are often multitrunked and can reach 70 feet (21.7 m) in height. The better specimens grow to be magnificent, looming trees covered with shining green, sharp needles glinting in the sun. As the tree matures, gray-brown outer bark peels off in strips to reveal attractive red inner bark. The squarish brown cones are unique and add an intriguing element to the tree. In China, this tree is also valued for its handsome, durable, and exceptionally workable wood.

China fir is hardy through zone 7 and it can be grown in protected sites in colder areas, but if the temperature drops to the range of −15 to −20°F (−23.5 to −26°C), foliage will be killed. In exposed, windy sites, or in regions with severe winters, China fir can suffer wind and cold damage resulting in significant dieback, which contributes to the tree becoming ragged with age. China fir prefers moist, well-drained, acid soil but performs well in wet, heavy clay soils. Beautiful large specimens can be seen in older neighborhoods across the southeastern United States. There are no significant pest or disease problems with China fir.

China fir is propagated from cuttings taken in winter. The cuttings are treated with very concentrated rooting promoters and rooted under mist. Production of a quality China fir takes several years. This is because young plants grown from cuttings taken from any branches other than terminals retain the lateral orientation of their parent branch and sprawl about until they reach a certain age. After three to five years of development, they begin to grow upright and form a normal, central leader. A block of young China fir in the field, branches lazing in all directions, is a bizarre sight indeed.

'Glauca' (Plate 67) is a lovely cultivar of China fir with waxy blue foliage. It is beautiful as both a young and mature tree. 'Glauca' may also have somewhat improved hardiness compared to the species. A new cultivar, 'Chason's Gift', from Johnson Nursery Corporation in North Carolina, is another excellent addition. The dark green, thickly foliated 'Chason's Gift' is uniquely pyramidal. It grows a strong single leader even as a young plant and retains a dense, pyramidal habit throughout its life.

The unusual foliage and pendulous branching of China fir lend it an air of exotica. The bold, glossy foliage, beautiful habit, and exotic character cannot help but catch the eye of each visitor to a garden where China fir resides. It is easy to see why the Victorians were so fascinated by the extraordinary China fir. It makes an excellent specimen tree in a site where it can be fully appreciated. A magnificent, multitrunked, 50-foot (15.2-m) specimen stands tall in the Raleigh Little Theatre Rose Garden in Raleigh, North Carolina, where its dramatic foliage and height are an exotic anchor for the unique sunken garden and a perfect balance to the more ethereal attraction of the roses.

Cyrilla racemiflora / Leatherwood   PLATE 68

The architectural qualities of ornamental plants are often the least advertised. But gardens rely upon the architectural characteristics of plants as much as any colorful displays they provide. The garden sculpture of weeping or contorted branches, or clean lines and skyward spires, are great gifts in the garden, representing impeccable architecture of the highest order. Finding manageably sized, architectural and ornamental plants that also perform well in low-maintenance landscapes is a challenge. While there are a multitude of such plants in the world, they are not among the plants getting the hard sell at most garden centers. Why? Because the customer is always in the garden center in the spring, picking out azaleas. We must start asking garden centers and nurseries for plants that offer architectural quality as well as showiness, and we can make an excellent start by asking for *Cyrilla racemiflora*, leatherwood.

Leatherwood is a deciduous to semievergreen large shrub or small tree with a loose, rounded habit and fascinating, open, spreading, twisted branching. The branches spread up and away from the trunk with a great deal of irregular twisting and contortion, as if sculpted by hand. The plant will generally reach 12–20 feet (3.7–6.2 m) in height with a nearly equal spread.

In addition to its architectural form, *Cyrilla racemiflora* offers interesting, delicately beautiful flowers in early summer. Pendulous clusters of small, creamy flowers are borne in rings around the base of the current year's new vegetative growth. The appealing whorls of flowers dangle lightly like the lace of demure wedding veils.

The evergreen character of leatherwood's leaves depends upon the degree of cold to which it is exposed. The plant itself is completely hardy through zone 5, but evergreen character will vary regionally. It may hold its leaves through the winter in warmer areas (zone 8 and milder), but will likely lose them in zone 7 and colder, and may even suffer some nonlethal winter damage in severe winters. But it is no loss if *Cyrilla* loses its foliage, because, in the process, it develops some of the best fall color to be found. The small, oval leaves turn brilliant canary-yellow, marbled with scarlet and orange.

Leatherwood is native to the swamps and wet places of the southeastern United

States and farther south into South America, but in the landscape it is tolerant of a range of conditions. It grows best in moist, well-drained, acid soils but also thrives in clay soils through droughts and wet years. Once established, it is an exceptionally tough plant. It does best in full sun or light shade and has no significant pest or disease problems. Leatherwood is readily propagated from summer hardwood cuttings treated with rooting promoter and rooted under mist, or from directly sown seed. It is most likely to be found through a nursery or catalog specializing in native plants.

The exceptional garden character of _Cyrilla racemiflora_ combines sculptural, floral, and foliar displays in a single plant. A specimen of leatherwood in the Southall garden of the North Carolina State University Arboretum shows the unparalleled work of nature's architects in this outstanding plant. As a player in a mixed border, a specimen with room to spread, or a star in a walled garden, leatherwood can bring the essential qualities of architecture to almost any garden.

Euscaphis japonica / Euscaphis   PLATES 69–71

At the North Carolina State University Arboretum in Raleigh, North Carolina, plants are evaluated for their beauty and adaptability, with the goal of making new and excellent landscape plants available to nursery and horticultural professionals, and thus to gardens. Hundreds of new plants are acquired each year from sources around the world to add to the arboretum's gardens for evaluation. New plants can come from seed exchanges, nurseries, other arboreta and botanic gardens, and, perhaps most excitingly, from plant collection expeditions to other parts of the world.

In 1985, J. C. Raulston was part of the United States National Arboretum collection expedition team that traveled to Korea. Cuttings and seeds of hundreds of exciting plants came back from that expedition, and many interesting plants have matured and are proving their mettle for production in the trade. Over the next several years, some of these plants will become available to gardeners—and they are plants well worth waiting and watching for.

One of the most unusual and delightful of the plants to come back from the 1985 Korean expedition is _Euscaphis japonica_, a plant heretofore so little known in the United States that it has no common name in English, although it is sometimes referred to by its fans as "Korean sweetheart tree," the name bestowed upon it by renowned nurseryman and plantsman Don Shadow.

Euscaphis is a small deciduous tree with thick, glossy, emerald-green foliage. It will likely reach 20 feet (6.2 m) with age, with a spread of about 10–15 feet (3.1–4.5 m) (remember this is a new plant). The medium-sized, oval leaflets are arranged in horizontal ranks, which gives the tree a neat appearance. The bark on the best of the seedlings is a dramatic violet-chocolate veined with bright white stripes and is unbelievably eye-catching, especially in winter (Plate 71). Small yellowish flowers are borne in early summer in branched clusters and are essentially unremarkable—but the fruit they develop into is another matter entirely.

In late summer, large clusters of leathery heart-shaped sacs develop, turning an intense cherry-red. As the season begins to turn toward fall, these romantic red hearts split open to reveal shiny, jet-black beads, creating one of the showiest fruit displays any gardener could wish for. Each tree appears bedecked with bouquets of brilliant valentines that positively glow against the rich green foliage—a surprising and wonderful effect in the late summer landscape.

Euscaphis has proven itself to be remarkably tough, growing and thriving through droughts and wet periods in some of the worst clay soils the southeastern Piedmont can dish up. Exact cold hardiness is as yet unknown, but the tree will likely perform well throughout the Southeast and most likely through zone 6, except in the most extreme areas of the mountains where cold may result in damage. Full sun will give best flowering and fruiting.

One reason why this plant has not been available has been the challenge of propagation. The seed requires multiple, varied treatments before it will successfully germinate. Now that these techniques have been worked out, *Euscaphis* can be propagated by nursery professionals and begin to work its way through the trade to gardeners and landscapers. In the meantime, *Euscaphis japonica* is a good reason to treat yourself to a trip to the North Carolina State University Arboretum, where this delightful tree wears its hearts on its leaves for all to see.

Feijoa sellowiana / Guava   PLATE 72

Is guava a tropical fruit or a unique landscape ornamental? It is both. Guava, *Feijoa sellowiana*, is a South American native that does indeed produce the luscious guava fruit, but it is also an exotically appealing shrub whose dramatic flowers could be the subject of a Georgia O'Keefe painting.

Guava is an evergreen shrub that will reach 10–15 feet (3.1–4.5 m) in height with an equal spread in tropical areas, but in warm temperate zones, it is more often seen at 5–6 feet (1.5–1.8 m) with equal spread. The shrub's habit is relatively dense in colder areas where late frosts and severe cold nip branch tips in the bud, but it can grow to be open and spreading in warmer areas where nature relies on horticulturists to do the pruning. The 2-inch (5-cm) leaves are covered with tiny hairs that give the foliage a silvery appearance.

Guava is a handsome evergreen shrub with unique character that would warrant a place in many gardens even if it never flowered, but the blooms of this plant are such treasures that they steal the show from the plant's other attributes.

The flowers of *Feijoa* are about 2 inches (5 cm) wide, and their coloring nearly defies description. A ring of snowy white sepals arches gracefully back and down to reveal velvety magenta petals surrounding an upright tuft of cherry-red stamens that are each dusted with bright gold pollen. The petal-like sepals are tasty to eat, right off the plant. The blooms are tucked throughout the plant among the foliage. It is always a special surprise when the first guava flower opens each year in late spring.

The flowers develop into green fruit that mature into yellow berries, 2–4 inches (5–10.1 cm) long, with a lush, pineapple-like taste. Fruit set is variable, with certain cultivars selected for fruit yield giving optimal fruit production. Fruit set may require, and is definitely optimized by, placing several plants in relatively close proximity to each other.

Feijoa sellowiana is reliably cold-hardy only to zone 8, but it can be grown in sheltered locations throughout the lower elevations of zones 7 and warmer. It grows best in moist, well-drained soil but will tolerate a range of less than ideal soils, from sand to heavy clays. Full sun gives the best flowering and fruiting. Guava will also perform well in partial shade (albeit with somewhat reduced flowering and fruit set). This shrub is quite salt-tolerant, and its silvery leaves and colorful flowers are particularly attractive in coastal landscapes.

Guava makes a dramatic specimen for a large container on a patio or near a walk. It can be pruned after flowering to control its spread. Landscape plants have been known to survive and reemerge after being killed to the ground by cold.

In addition to seedlings, which are generally good plants, there are several cultivars of _Feijoa sellowiana_, which have been selected primarily for fruit production.
'Coolidge', 'Nazemata', and 'Pineapple Gem' are all good self-pollinating selections.
'Superba' is a round-fruited form that needs to be planted with another cultivar for
good fruit set. 'Variegata' has white-variegated foliage. Propagation of guava is most
reliable from seed separated from the ripe fruit, but cuttings taken in summer are
rooted for propagation of cultivars.

    Feijoa sellowiana never fails to command attention in the landscape. As visitors en-
counter this spectacular plant as they leave the white garden at the North Carolina
State University Arboretum, they are always surprised to learn that it is the parent of
a familiar fruit. The unique guava—landscape ornamental and tropical fruit—deserves
much wider use in mild-winter gardens.

Firmiana simplex / Chinese parasol tree   PLATE 73

    Chinese parasol tree adds an exotic flavor to the landscape with its smooth green
bark, its wide, lush leaves, and its showy, distinctive "parasol" fruit capsules. This fascinating tree hails from the more temperate areas of China and Japan—areas more similar in climate to the southeastern United States than the true tropics of the world.

    Firmiana simplex is a large deciduous tree, 20–40 feet (6.2–12.4 m) in height, which de-
velops a bold-textured, rounded canopy. The remarkable leaves are exceptionally
large, often as long as a foot (0.3 m) and nearly as wide, somewhat maplelike in shape,
and a deep grass green. Foliage may develop a bright canary yellow fall color but this
character is not always consistent. The tree is hardy to zone 7, making it a useful addition to landscapes or gardens where a tropical effect is desired (imagine it as a shade
tree near a walkway or patio at a summer retreat home).

    Firmiana simplex blooms in early to midsummer with long clusters of small, yellow-
green flowers. Rather inconspicuous individually, the blossoms are clustered in showy,
2-foot (0.6-m) panicles that arch up and out from the foliage at the ends of the branches.
The bark is beautiful—a surprisingly smooth, bright moss-green.

    In late summer, the canopy is bedecked with hundreds of miniature "parasols" dan-
gling overhead. These parasols are actually the papery outer skins of the fruit, which
separate and pull away from the seed inside but remain draped like tiny umbrellas
from the fruit stem around the seed. In late summer and early fall, the parasols are a
fresh cream color tinged with pink and green, but as they age and dry, the capsules
turn a warm cinnamon-tan. The dried parasols are excellent additions to arrangements
of everlastings.

    Firmiana simplex is one of the trees brought from Asia in the early era of plant col-
lecting in the late 1700s. It performs well in a range of soils, including very heavy clays.
It will tolerate part shade but needs full sun for best canopy development, flowering,
and fruiting. Chinese parasol tree is propagated from seed, which germinates easily
without any special treatment. It is not commonly available but may be seen in arboreta and botanic gardens and can be purchased from a few specialty mail-order nurseries. At the North Carolina State University Arboretum, you can find Chinese parasol
tree across from the _Stewartia_ collection at the west entrance to the Winter Garden.

    Chinese parasol tree brings a stroke of the tropics to gardens in need of a bold hand.
The smooth green bark makes an unusual foil for variegated climbing vines—another
way to create an element of tropical ambience. If you half-close your eyes as you stand
underneath and gaze up at the little parasols, with the thick air of late summer cling-

ing around you, you can almost make out those long-gone Victorians, fanning themselves and twirling their parasols under a far-off tropical sun.

Gordonia lasianthus / Loblolly bay   PLATE 74

A heady fragrance drifts on the humid air of summer from the creamy blossoms of loblolly bay, an evergreen tree with leaves like camellias and wonderfully exotic blooms. A member of the tea family, which also includes the well-known camellias native to Asia, *Gordonia lasianthus* (Plate 74) is an American native that shares the exotic blossoms and shiny foliage of its better-known relatives. Loblolly bay is also a close cousin of the fascinating *Franklinia alatamaha*, a tree that was discovered in the late 1700s along the banks of Georgia's Altamaha River but which has not been seen in the wild since 1790, although it is successfully carried on in cultivation.

Loblolly bay is a broad-leaved evergreen tree that reaches 30–40 feet (9–12.4 m) in cultivation. It sometimes attains greater height in the wild. An open, somewhat gangling tree when young, it develops a pyramidal, more dense appearance at maturity. Foliage is a rich green, similar to evergreen magnolias, which creates a beautiful setting for the large white flowers. The alluring flowers are fully 3 inches (7.6 cm) across, with a ring of golden yellow stamens encircling the center of the bloom. Clusters of several buds are borne in the leaf axils, and in midsummer, the flowers open, one at a time in each cluster. The flowers mature to tannish, dry capsules, which split open in the fall. The dark gray bark of loblolly bay, once used to tan leather, becomes attractively fissured as the tree ages (a distinguishing character from the smoother bark of *Franklinia alatamaha*).

Loblolly bay is hardy to zone 8. Native to the Coastal Plain and Piedmont of the American Southeast, the tree is often found growing with sweetbay (*Magnolia virginiana*) in the moist soils of bays and swamp edges. While it prefers moist and acidic landscape conditions, in cultivation it requires a well-drained site. Attempts to plant loblolly bay in "swamp" conditions created by severely compacted clays in housing developments will probably kill this plant. Wet areas in the wild where loblolly bay grows are usually wet only during times when the plants growing in them are dormant and so require little oxygen or nutrients to survive. Planting in permanently wet areas, or areas that are repeatedly drowned during the growing season, is asking too much of their native tolerances for periodic flooding and wet feet. Nonetheless, loblolly bay is not as demanding as it sounds. Given the appropriate site considerations of well-drained, loamy soils, it will perform well in hot, humid climates, but it is not tolerant of poorly drained soils. Full sun or partial shade will give good growth. Flowering is best in full sun, but part shade will prevent winter burn on the foliage.

There are no special cultivars or selections of loblolly bay readily available, but the native species can be found at specialty nurseries, especially those focusing on native plants. It is relatively easy to propagate from cuttings or seed. Summer softwood cuttings root readily under mist, even without rooting promoter. Seed is collected by harvesting the pods before they split open. Pods contain upwards of 100 tiny seeds, which must be extricated, soaked in lukewarm water for two days, air-dried, and sown in a light medium for germination, which can take up to two months.

The loblolly bay is one more example of just how magnificent our native flora can be. Its wild, native beauty adds an unusual element to gardens with organic, well-drained soil. It is very effective as a specimen in small gardens, especially when underplanted with bulbs, and it combines well with other evergreen trees and shrubs, which set off its lovely flowers.

Hedera helix / English ivy   PLATE 75

Hedera helix, English ivy, is a well-known evergreen, woody vine that is either well loved or well hated for its indomitable ability to rampage across the landscape regardless of the challenges. But there are hundreds of cultivars of English ivy, and many of these named forms are significantly less rampant growers and far better behaved choices for small gardens. English ivies are excellent choices as both groundcovers and climbers. Selected for differing leaf size and shape, the cultivars offer a lovely range of variegations, including creams, whites, golds and even pinks. Few things lighten up the dark corners of a wooded garden as well as a white- or gold-variegated ivy slipping up the tree trunks.

If you have had the pleasure of training English ivy up a high wall, tree trunk, post, or fence, you may have noticed an interesting transformation that occurs as the vine grows into the upper reaches of the vertical. The pronounced, sharply three-lobed shape of the leaf begins to change into a narrow heartlike shape. You may have also noticed the emergence of small, starry flower clusters late in the summer that punctuate the upper reaches of the vine with haloes of gold, eventually maturing into handsome (but poisonous!) inky black fruit. What you have seen is the rather amazing transformation of the vine from its juvenile to its adult physiological state. This fascinating change in character is triggered by the vine growing a certain vertical distance from its roots.

Adult *Hedera helix* offers different landscape character than its youthful form. The adult leaf shape is less bold and more refined than the juvenile leaves. The flowers and fruit are subtly attractive in late summer, especially against the leaves of variegated forms. Once a vine attains the adult form, it can be propagated and grown so that the propagated plants start out as adult-form plants. Specialty nurseries sell adult forms. Adult forms tend to develop an attractive shrubby habit, but they must remain upright in order to retain their adult habit. If planted at ground level with no vertical surface nearby, they will quickly revert to the juvenile form with sharply lobed foliage and no flower or fruit production.

Adult-form plants are beautiful in the corners of walled gardens and other small spaces. In the Victorian era in England, many cultivars of adult ivy were produced and available commercially, but they are available only from a few specialty nurseries today. Old specimens of globose ivy shrubs, 6–10 feet (1.8–3.1 m) in diameter, as well as true hedges of adult ivy, can be seen in old gardens and arboreta (an ivy hedge can be seen in the herb garden at the United States National Arboretum).

One of the advantages of adult ivy is that it is generally slower growing and more manageable than juvenile forms, while remaining tough and adaptable in a range of sites. *Hedera helix* can be completely hardy to zone 4, depending on the cultivar, but it needs to be in a protected site in the coldest parts of its hardiness range to avoid winter damage. It grows best in moist soils that are high in organic matter, but it will also thrive in most other soils, from wet clay to light sand, and it is quite salt-tolerant. Both adult and juvenile English ivy perform well in a range of light conditions, from full sun to dense shade. In areas with intense sunlight, the foliage of variegated forms is best in part or full shade. It can be propagated from cuttings almost any time of the year. Cuttings can be difficult to root but will root best if treated with rooting promoters and rooted under mist. As a result, adult ivy cultivars are sometimes propagated by grafting onto *Hedera helix* understock.

Adult forms of English ivy can often be seen in large trees around older, established gardens, but the plant can also add luster and refinement to difficult, tight urban gardens. Because of their shade tolerance, ivies are wonderful in corners of walled gardens. They are exquisite container specimens. Gardeners with a special interest in ivies of all types may want to contact the American Ivy Society, c/o William Redding, P.O. Box 520, West Carrollton, Ohio 45449-0520.

Hibiscus syriacus / Rose-of-Sharon   PLATE 76

Rose-of-Sharon, *Hibiscus syriacus*, is not a rose at all but rather a species of mallow. This old-fashioned shrub was a favorite of our grandparents and is often found in older gardens and around historic homes. A native of East Asia, the rose-of-Sharon became very popular in American gardens because of its large, soft-colored blooms and relatively trouble-free culture. In fact, it is so trouble-free that it is often found naturalized in semiwooded areas close to its original plantings.

Hibiscus syriacus thrives in hot climates. This deciduous plant generally grows as an erect shrub, 8 to 10 feet tall. Glossy green foliage is the backdrop for flowers with petals like crepe paper and soft, pastel colors. The blossoms vary in size, from 3–6 inches (7.6–15.2 cm) wide, depending on the cultivar, and the shrub blooms abundantly throughout most of the summer.

Rose-of-Sharon is reasonably tolerant of most sites, as long as they are not extremely dry or wet. Cold-hardy to zone 5, it performs best in moist, well-drained soil in full sun or partial shade. Since it blooms on new wood, a heavy pruning in winter will increase flowering the following spring. There is no appreciable fall color, but the persistent dry fruits offer winter interest.

The multitude of rose-of-Sharon cultivars, with flower colors ranging from white, pink, and red to lavender and blue-lilac, makes it easy to find one that pleases. 'Diana', a relatively recent introduction from the National Arboretum, is more compact than many rose-of-Sharon cultivars. It has dark green foliage and large, pure white blooms that remain open at night. Another interesting cultivar is 'Blue Bird' (Plate 76), with flowers that are nearly true blue in color. The cultivar 'Purpureus Variegatus' is especially interesting, as it offers white-variegated foliage.

Rose-of-Sharon is perfect for bringing gracious color to the summer landscape. Because it flowers through most of the porch season, it makes a good privacy screen for lazy afternoons in the porch swing. The romantic flowers framed by soft green foliage are the essence of an old-fashioned summer of a gentler era.

Hydrangea paniculata / Panicle hydrangea   PLATE 77

In the dog days of late summer, a gardener's eye is caught by the image of creamy white pyramids of blooms suspended above broad green leaves. These delightful floral crowns are not a mirage of summer's heat but the flower clusters of *Hydrangea paniculata*, panicle hydrangea.

Panicle hydrangea is a large deciduous shrub or small tree with rich green leaves that can reach anywhere from 10–20 feet (3.1–6.2 m) in height with equal spread. The shrub's natural habit is coarse and open, with somewhat loose, arching branching. If a tidier habit is desired, plants can be pruned in winter to encourage a tree form. This type of pruning has the added advantage of revealing the deep gray or brown bark, which becomes ridged as the plant matures. Since panicle hydrangea blooms on new wood, the plant can also be cut to the ground each year and treated as a woody herbaceous perennial, a good way to handle the plants where space is limited.

Each of the distinctive pyramidal clusters is made up of a combination of many small fertile flowers, which are not showy, as well as larger, showy sterile flowers. The proportion of sterile to fertile flowers will vary with cultivar and individual seedling plants. The flowers emerge a soft white, which turns slightly pink with age and eventually dries to brown. The panicles are held on the plant through the fall, a characteristic considered late-season interest by some but untidy and unattractive by others. Foliage offers little in the way of fall color.

Hydrangea paniculata is native to Asia and is one of the hardiest hydrangeas available, being reliable to zone 3. It prefers moist, rich, well-drained soils but also grows well in clay. Full sun is best for the most prolific flowering, but partial shade can also result in a good display. The plant is propagated by rooting spring softwood cuttings under mist.

Probably the most well-known cultivar is 'Grandiflora', peegee hydrangea, whose flowers are all sterile and exceptionally showy. 'Grandiflora' flowers from early summer on. Its flower clusters can be extremely large and heavy, weighing the branches down to the ground after a rainstorm. 'Floribunda' flowers later in the season. Its smaller panicles, with far fewer sterile flowers than 'Grandiflora' (but still more than the species), produce a more subtle display. 'Praecox' is an especially vigorous and early-flowering cultivar, which blooms in May or June in zones 6–7. 'Tardiva', on the other hand, is significantly later than other panicle hydrangeas. Most years, 'Tardiva' does not begin blooming until August or September in zones 6–7. It has a somewhat more upright habit than other cultivars. 'Unique' (Plate 77) is a lovely form from England with midseason flowering. A plant of 'Unique' is nestled in the mixed shrub border at the North Carolina State University Arboretum, illustrating a wonderful landscape use for panicle hydrangea, while 'Tardiva' contributes beautiful white color and dimension to the early-late border, at the west end of the perennial border, at the Arboretum.

Panicle hydrangea is a choice plant for a border of mixed shrubs, or as an informal mass. It adds late summer interest to the lawn or a wooded area. The cool white flowers brave the heat at a time when the landscape itself seems to lie back and rest in the shade.

Hydrangea quercifolia / Oakleaf hydrangea   PLATES 78, 79

Oakleaf hydrangea is a brash, bold-textured shrub whose leaves resemble those of a vigorous red oak, both in size and shape. These look-alike leaves add an unusual touch to the spreading, rounded shrub, as if the boughs of an oak had been laced into a rounded framework. In addition to the marvelous foliage, oakleaf hydrangea sends up foot-tall cones of creamy white flowers in early summer.

This large, multistemmed shrub can reach 5–8 feet (1.5–2.4 m) tall. The leaves emerge a downy gray color in the spring, folded upright in pairs like clasped hands. The flowers mature from white to rosy pink. The oaklike foliage stays green until late in the fall, when the leaves turn vivid purple, orange, and scarlet. Oakleaf hydrangea's warm brown bark peels and is very appealing both during leaf-fall and throughout the winter.

Hydrangea quercifolia is native to the Deep South, where it can be seen growing in moist sites, but it performs beautifully in other regions and is very well adapted to a great diversity of landscape conditions. It is hardy to zone 5 but may suffer some injury during severe winters if temperatures drop toward −10°F (−21°C). The flower buds will be killed when temperatures drop into this range, but the plant will survive and make an excellent foliage plant under those conditions. Oakleaf hydrangea prefers moist, fertile, well-drained soil but will tolerate more poorly drained conditions. Full sun or partial shade are best, especially if the roots can be kept cool. It does not do well in very dry areas. There are no serious pest or disease problems. The plant is readily propagated from fresh seed, but it is more difficult to propagate vegetatively. Cuttings taken in early summer can be rooted under mist in an exceptionally well-drained rooting medium.

There are quite a number of cultivars of oakleaf hydrangea, all of them superior to the species in both profusion of flowers and development of fall color (a very variable trait in seedlings). Some of the better known cultivars include 'PeeWee', a particularly compact form, reaching only 3–4 feet (0.9–1.2 m) in height with an equal spread; 'Snowflake', with nearly double flowers; 'Snow Queen', from Princeton Nursery in New Jersey, with exceptional fall color, very large florets, and upright flower clusters that droop minimally under their own weight; and 'Alice' and 'Alison', two selections that plantsman Michael Dirr has made for fall color and upright flowers.

Oakleaf hydrangea is a shrub for four-season interest in the garden. Its bold texture requires care in placement, so that it does not overpower the quieter beauty of other garden plants. Oakleaf hydrangea makes a strong statement in a mixed border where it can really hold its own, or in a massed planting for real drama, or as a focal plant in smaller gardens.

Hypericum spp. / St.-John's-worts   PLATES 80, 81

St.-John's-worts were highly valued in medieval England for use in a variety of folk medicines, but that doesn't mean they should be restricted to the herb garden. The variety of forms of today's hypericums can offer delightful beauty throughout the landscape.

The genus *Hypericum* includes species originating from a broad spectrum of climates stretching from Europe through the Mediterranean into China and other parts of Asia, and across the Atlantic to North America. A number of species and their hybrids and cultivars are lovely garden plants. They range in character from evergreen or deciduous medium-sized shrubs to groundcovers and herbaceous perennials, but they all have in common a display of bright, sunny yellow flowers like generous buttercups, which develop into warm brown capsule-type fruits in the fall.

One of the best hypericums for the garden is the native *Hypericum frondosum*, golden St.-John's-wort (Plates 80, 81). Hardy to zone 5, this deciduous shrub is covered with 1–2-inch (2.5–5 cm) diameter butter-yellow flowers in June and July. The flowers of golden St.-John's-wort are unique compared to its other *Hypericum* cousins because of the little tufted cushion formed in the center of the flower by the numerous, densely clustered stamens. Golden St.-John's-wort offers one of the best floral displays of the hypericums, but it also has other special traits. The foliage is a very refreshing blue-green that contrasts beautifully with the golden flowers. As the shrub reaches its mature height of 3–4 feet (0.9–1.2 m), the bark will peel to reveal an attractive red-brown color. 'Sunburst', a cultivar of *H. frondosum*, is somewhat more compact and lower growing than the species.

Named for Peter Kalm of Sweden, *Hypericum kalmianum* or Kalm's St.-John's-wort reaches 2–3 feet (0.6–0.9 m) high. It has a loose and open habit, with lemon-yellow flowers that bloom in early July. It is hardy to zone 4 and may be evergreen in protected areas of zone 8 and warmer. Kalm's St.-John's-wort has somewhat fragrant flowers, and the leaves are whitish on the undersides.

Hypericum patulum, golden cup St.-John's-wort, is another attractive shrub form of St.-John's-wort. It is somewhat shorter than the others, reaching a maximum of 3 feet (0.9 m), and has semievergreen to evergreen foliage in zones 7 and warmer.

Other good shrub forms of _Hypericum_ include the hybrids _H._ 'Hidcote' and _H._ × _moseranum_, Moser's St.-John's-Wort. 'Hidcote' is hardy to zone 6. It reaches about 3 feet (0.9 m) in height and sports very dark green foliage. The flowers are somewhat larger than those of golden St.-John's-wort, but they do not have the latter's delightful central tuft, and they are more orange-gold in color than the clear butter-yellow of _Hypericum frondosum_. Moser's St.-John's-wort is a hybrid bred in Moser's Nursery of Versailles, France, in the late 1800s. It is a compact, rounded shrub, reaching 2–3 feet (0.6–0.9 m) in height with a spread of about 3 ft (0.9 m). The foliage is blue-green and the golden yellow flowers are quite large, 2–3 inches (5–7.6 cm) inches across, with interesting pinkish purple heads on the stamens. It flowers late in the season, from July through fall in zones 6–8. _Hypericum_ × _moseranum_ is not as hardy as the other hypericums, being reliable only to zone 8 (to zone 7 with protection). In zones 5 and 6, it should be treated as a herbaceous perennial and cut back to the ground each winter. 'Tricolor' is a white- and rose-variegated cultivar of Moser's St.-John's-wort, grown at Wisley Gardens in England; while the variegation is interesting, the habit of the plant is somewhat sparse and rangy.

Aaron's beard St.-John's-wort, _Hypericum calycinum_, is a woody groundcover form that grows only 1–1.5 feet (0.3–0.4 m) tall but spreads to 2 feet (0.6 m). The large flowers, 3 inches (7.6 cm) wide, are a bright, clear yellow set against dark green foliage. This lovely groundcover deserves greater attention in the landscape. It is hardy to zone 5. While it may suffer some winter dieback in colder areas, flowers are formed on new wood so the flower display should remain unharmed. One way to handle this plant is to cut it back to the ground in late winter each year to stimulate a new flush of growth. Since growth is rapid, recovery and display will be improved.

Hypericum forrestii, a relatively new _Hypericum_, is a deciduous shrub with golden yellow flowers and smaller, more finely textured foliage than that of the other hypericums. It reaches 3–4 feet (0.9–1.2 m) tall. _Hypericum forrestii_ sometimes has handsome wine-red fall color when conditions are right; fall color also appears to vary with seedlings. The plant is being tested and evaluated by professionals in the field, and is expected to be commercially available in the next few years. It has not been planted out in trial at the North Carolina State University Arboretum, but has been distributed through the Arboretum's introduction program for testing and evaluation.

Culture of _Hypericum_ varies a bit depending on the species or cultivar you are growing. All St.-John's-worts prefer dry soils in full sun to partial shade, but many will also perform well in clay soils if placed in full sun. Best choices for long-term performance in heavy soils include _H. frondosum_, _H. calycinum_, and _H. forrestii_. Many of the other St.-John's-worts may ultimately succumb to hot, wet summers in heavy soil after two years or so. The best choice for northerly areas with cold winters is _H. kalmianum_, the most cold-hardy of the hypericums. St.-John's-worts are easily propagated by cuttings or by seed. Softwood cuttings taken in the summer are rooted under mist in a light medium, such as sand. Seed requires no special treatments.

The incredible array of St.-John's-worts offers a delightful palette of sunburst golds for summer gardens. They are excellent as massed groupings along drives and walls, or as special accents in mixed shrub borders and small-scale landscapes.

Ilex vomitoria / Yaupon   PLATE 82

Ilex vomitoria is named for its medicinal use as a cathartic purge, a native American remedy that used the leaves as a tea. In the wild, this evergreen large shrub or tree grows in wet areas throughout the mid-Atlantic and southeastern regions of the United States, from New York to northern Florida and west to parts of Texas. It has been in cultivation for some time and has proven itself to be a tough, attractive evergreen for the mid-Atlantic and south.

In cultivation, yaupon is generally an upright small tree with irregular, somewhat picturesque branching (depending on cultivar). Growth rate varies from medium to fast, depending on the form and conditions. Full-size forms reach 12–20 feet (3.7–6.2 m) in height with a spread of 8–12 feet (2.4–3.7 m). The leaves are dark green, neat and small, reminiscent of those of boxwood (*Buxus*) or Japanese holly (*I. crenata*), for which it is often used as an alternative. The habit of yaupon is quite different from boxwood (with the exception of the cultivar 'Nana'), but the habit of the compact forms of yaupon is close to that of many Japanese hollies.

The evergreen foliage of yaupon is reliably attractive throughout the year, with none of the pest or disease problems exhibited by other plants used in the same landscape niche, such as Japanese holly. It produces brightly colored red or yellow fruit (depending on the cultivar). Fruit is produced only on the female plants but is generally very abundant when males are also growing in the neighborhood.

Ilex vomitoria is reliably hardy through zone 7 and in sheltered sites into zone 6. It is tolerant of a broad range of soil conditions, from wet to droughty. The plant does best in full sun but also performs well in partial shade, with the compact forms being most tolerant of shade. Yaupon is propagated by cuttings, which are taken in early fall or late spring, treated with rooting promoter, and rooted under mist.

The multitude of cultivars makes it possible to use this excellent plant in many ways, from unobtrusive but dependable foundation plant (*Ilex vomitoria* 'Nana') to exquisite specimen (*I. vomitoria* 'Pendula'). Yaupon takes well to clipping and pruning, and the taller forms make good barrier hedges and screens.

Many fine cultivars of *Ilex vomitoria* are available. 'Folsom's Weeping' is a tall, narrow, strongly weeping female; 'Grey's Little Leaf', a delicate, lacy dwarf with leaves half the size of the species; 'Jewel', a rounded, compact form with heavy red fruit production. 'Nana' is a compact form that eventually forms a broad cushion to 5 feet (1.5 m) in height. It is a popular and excellent substitute for Japanese holly. 'Pendula' (Plate 82) (also listed as *I. vomitoria* f. *pendula*) is a tall, weeping form that is found as both male and female, with the female bearing abundant cherry-red fruit the size of small pearls. 'Shadow's Female' is a large shrub form with exceptionally dark green foliage and scarlet fruit that seems to have improved cold hardiness. 'Straughn's' is a compact form with the informal mounded habit of boxwood, for which it is a good substitute, although it tends to open up in the center with age. 'Yellow Berry' is a good yellow-fruited form; 'Wiggins Yellow' bears bright yellow fruit. 'Will Fleming' is an unusual yaupon with formal, narrowly fastigate habit and gray-green foliage that resembles a small Italian cypress (*Cupressus sempervirens*) from afar.

At the North Carolina State University Arboretum, two plants of the elegant 'Will Fleming' flank the entrance to the visitors' center. Visitors are welcomed to the gardens there by these holly gentlemen, illustrating how they can bring panache to your garden.

Indigofera spp. / Indigos   PLATE 83

Sources of fresh floral color in the garden are few and far between in late summer, before the fall extravaganza. The relatively small number of shrubs and trees that do bloom at this time are greatly valued for their "tardy" floral display. One group of plants that is always welcome for late-summer bloom is _Indigofera_—the shrubby indigos. All _Indigofera_ in cultivation in the United States are leguminous shrubs reaching 2–6 feet (0.6–1.8 m) in height, depending on species, with an equal to broader spread. They are somewhat reminiscent of the _Lespedeza_ species (shrub bush clover) in habit but have a more refined texture and flower. The divided foliage, shaped like that of other pea-family members, is a fresh light green, even in August, and is held on upright, gently arching branches emerging from the crown. The flowers are a warm rose color and brighten the ends of the branches with color that is a standout in the late summer border.

Of the roughly 700 species of _Indigofera_ found around the world, only a relative few are in cultivation in America. The most famous is _I. tinctoria_, the source of the deep blue indigo dye in use by human cultures since 2300 B.C. The species differ slightly in cold hardiness, form, and flower color, but all thrive in full sun in a range of soil conditions from light sand to heavy clay. There are no significant pest or disease problems. Severe winters may kill branches to the ground, but the plant will regrow from the crown. Seed is most commonly used for propagation; softwood cuttings rooted under mist may be possible with some selections.

Among the species is Himalayan indigo, _Indigofera heterantha_ (sometimes seen as _I. gerardiana_), which reaches 2–6 feet (0.6–1.8 m) in height and bears deep rose flowers. It is reliably hardy to zone 6 and possibly colder areas. _Indigofera incarnata_, Chinese indigo, is much lower growing than other species, reaching 1–2 feet (0.6–0.9 m) in height. It has dark green foliage and bright pink flowers held in very long racemes. The botanical variety of Chinese indigo, _I. incarnata_ var. _alba_, produces creamy white flowers and is reportedly more cold-hardy than the species. Another beautiful white indigo is _I. decora_ var. _alba_, which is a very low growing form, reaching only 10–12 inches (25–30.4 cm) in height. _Indigofera kirilowii_, Kirilow indigo, is native to Japan, Korea, and China. It is a relatively low shrub, reaching 3 feet (0.9 m) in height and is hardy to zone 5 and possibly colder. Kirilow indigo's flowers are medium rose in color and are not quite as showy as others because they are more hidden among the foliage. But _I. kirilowii_ has lovely gold fall color that is wonderful in combination with other shrubs, and perennials, in the fall border. Horticultural hybrids include _Indigofera_ 'Rose Carpet', a low form with bright rose flowers.

None of these plants are likely to turn up in large-volume retail sources, but with some hunting through mail-order nurseries that work with new and unusual plants, _Indigofera_ can be found. The jewel-toned flowers of _Indigofera_ combine easily with many perennials and grasses, making this informal shrub a premier choice for the border. It also makes an effective screen when used as a mass for landscape eyesores such as utility meters, or for seasonal screening around pools and patios.

Itea spp. / Sweetspire   PLATE 84

Ropes of tiny pearls dangle at the ends of branch tips, offering a subtly spicy fragrance carried on the warm breeze of early summer. These pearly treasures are the flowers of Virginia sweetspire, _Itea virginica_, a deciduous to semievergreen shrub, reaching 3–6 feet (0.9–1.8 m) in height with an arching, beautifully rounded habit. The branching is upright, but as the plant reaches maturity, the branches tend to arch over, with their flower clusters draped toward the ground like a fountain.

Virginia sweetspire is native to the wet places and pine barrens of the East Coast, from New Jersey to Florida and west to Louisiana. The flowers are in full bloom in early summer and add an understated touch of white to the garden. The foliage of Virginia sweetspire is a very attractive dark green. In fall, it turns to cardinal red, wine, and purple. *Itea virginica* is hardy to zone 5. In mild winters, above 15°F (–8.5°C), much of the foliage on the plant will persist through the winter in subdued tones of its fall color and in olive-green.

Virginia sweetspire is a fine low-growing plant for group planting, spreading to form masses. It can be either invasive or expansive, depending on its place in the landscape and the control tactics of the gardener. To keep a mass in check, root prune each year about 1 foot (0.3 meters) out from the dripline. *Itea virginica* can be grown in conditions ranging from full sun to shade. In the wild, it prefers deep, moist, loamy soils, but in cultivation it has proven to be drought-tolerant and quite a tough, stress-resistant plant.

Cultivars of Virginia sweetspire include the award-winning 'Henry's Garnet' (Plate 84), released from the Scott Arboretum in Swarthmore, Pennsylvania; 'Saturnalia' from planstman Larry Lowman in Arkansas; and 'Sarah's Eve', introduced in 1990 from Woodlander's Nursery in Aiken, South Carolina. 'Henry's Garnet' has exceptional garnet-red fall color and very large flower clusters that can be 6 inches (15.2 cm) long. The fall foliage of 'Saturnalia' is a vivid mix of brilliant oranges, purples, and wine. The petals of 'Sarah's Eve' have a slight pink blush that disappears at full flower, and the pedicels (and other smaller flower parts) are colored a deep wine-red.

Another deciduous species, Japanese sweetspire, *Itea japonica*, is native to Japan and grows rapidly to reach a mature height of about 3–4 feet (0.9–1.2 m). It is not as hardy as the native *Itea virginica* and likely not reliable in regions colder than zone 6. The National Arboretum released a cultivar of Japanese sweetspire called 'Beppu', which has exceptional foliage and good, dense flowering displays.

Two other sweetspires offer evergreen beauty. Holly sweetspire, *Itea ilicifolia*, has dark, glossy green leaves, which resemble holly leaves only in that they are somewhat spiny on the edges and persist all winter. Holly sweetspire is taller than *Itea virginica*, reaching 6 to 12 feet (1.8 to 3.7 m), and it bears long chains of greenish flowers. Because of its evergreen character, holly sweetspire is less stress-resistant and needs some shade and good moisture. It is reliably hardy only to zone 8. Chinese sweetspire, *Itea chinensis*, is proving to be a much hardier and more vigorous evergreen shrub in trials at the North Carolina State University Arboretum in Raleigh, North Carolina. It has long, narrow leaves and will apparently reach 10 feet (3.1 m) in height. Chinese sweetspire is not yet in commercial production in the United States, but appears to hold potential for use as a screening plant. It blooms somewhat earlier than other *Itea*, starting in April in zone 7. All of the sweetspires are easily propagated from softwood cuttings rooted under mist.

At the North Carolina State University Arboretum, sweetspires make a soft and lovely foil for other plants in the mixed shrub border. The plants bring a pearly lustre to any border's early season glow of bloom. Plant these exquisite shrubs in your own mixed border to discover that sweetspires can be priceless gems.

Kadsura japonica / Kadsura vine   PLATE 85

August brings ennui to the landscape. Fall has not yet arrived, but the season has taken a noticeable turn away from active growth and toward the hardening of autumn. To remedy this, landscapes often rely on bright perennials and bedding plants to shine through the end of summer. Many gardens, however, are not well suited for this blaze of color, yet they would still benefit from late summer freshness that is not dependent on such bold splashes. Kadsura vine is one of the best plants to bring visual vigor to the garden at this time of year, without depending on intense color.

Kadsura japonica is an evergreen, twining vine, native to Japan, that is well adapted to landscape conditions of areas with hot summers and heavy clay soils. The formal, oval leaves, which reach 4 inches (10.1 cm) long, are very dark green and highly glossy. Because of this high-gloss finish, the foliage is bright and fresh in appearance, even through late summer. Leaves are suspended from the main vine stem with cherry-red petioles, which also contribute to the bright character of the plant. The evergreen foliage remains good-looking through mild winters. Kadsura flowers in midsummer with small yellow blooms that mature into scarlet fruits that show vividly against the dark foliage. The fruits persists through the fall and into early winter.

Kadsura is quite an elegant vine. Unlike some other vines, it is very well-behaved and will not surge over all vertical surfaces within reach. It does grow vigorously, however, and is an excellent choice for arbors, posts, or any other surface requiring the grace of an evergreen vine. Kadsura is hardy to zone 8. It can be grown in a range of conditions but does best in partial shade with good moisture. It tolerates heavy soils and full sun but may suffer some foliar scorch in the winter in full sun. There are no disease or pest problems with this vine.

A few cultivars are available, although somewhat difficult to find. 'Shiromi' is a white-fruited form, while 'Chirimen' and 'Fukurin' (Plate 85) have varying degrees of cream-colored variegation on the foliage. The variegated forms are particularly striking. 'Fukurin', the form with greater variegation, is especially useful for brightening the garden. Kadsura is propagated from cuttings taken in summer and rooted under mist.

At the North Carolina State University Arboretum, kadsura vine greets visitors as they enter the white garden. The vine keeps an enticing freshness for the entire summer. This freshness of foliage on a climbing vine can be used in many ways: on an arbor or trellis, climbing up a fence or bird feeder, twining along a handrail, or softening an iron fence. Kadsura vine shines glossy and untarnished throughout the summer.

Koelreuteria spp. / Golden-rain tree   PLATE 86

Golden-rain tree is a perfectly named plant. Large clusters of bright golden yellow flowers wave over the light green foliage, then, after blooming, the flowers fall to the ground as a golden rain, creating a lovely carpet around the base of the tree. The flowering display is followed by dry, papery fruits reminiscent of tiny Japanese lanterns.

Two species of golden-rain tree are commonly planted in the landscape: *Koelreuteria paniculata*, golden-rain tree, and *Koelreuteria bipinnata*, bougainvillea golden-rain tree. Both are medium-sized trees with clusters of golden flowers and light green, compound foliage. All *Koelreuteria* species are adaptable to most soils, as long as the site is reasonably well-drained and in full sun.

Golden-rain tree is the most common and readily available. It is a spreading tree reaching 30 feet (9 m) in height. Its foliage is more divided than that of bougainvillea golden-rain tree. It flowers in early summer, and its seed capsules are light brown. Hardy to zone 5, *Koelreuteria paniculata* is the choice for northern gardens.

Bougainvillea golden-rain tree (Plate 86), hardy to zone 7, is the preferred species for zones 7 and warmer with long, hot summers, because it tolerates the heat. It flowers in early autumn, later than its relative. The papery capsules are a delightful rosy pink color, and the fernlike foliage turns a lovely clear yellow in the fall. A native of China, *Koelreuteria bipinnata* was introduced to American gardens in the late 1800s. It is a fast-growing, multistemmed tree that will reach 25–30 feet (7.5–9 m), with a pleasingly upright and spreading form. It is advisable to prune young trees to achieve a full, dense crown at maturity. Bougainvillea golden-rain tree will flower at two to three years old. Planted in full sun, it will flower profusely once it begins.

There are some challenges in obtaining bougainvillea golden-rain tree. Not only is it difficult to locate a producer, but there is confusion among growers as to which species of *Koelreuteria* they are actually working with. This is because the bougainvillea golden-rain tree is essentially identical in appearance to a third species, *Koelreuteria elegans*, flamegold, which is not hardy beyond zone 9. The only way to tell the two apart is to see which survives the winter! Since the source seed from Asia is often misnamed, nursery owners must rely on the plant's performance for identification. To insure that the plant in your garden is truly bougainvillea golden-rain tree and not flamegold, buy it from a nursery in zone 7 or colder, where the grower has been able to overwinter the plant successfully.

Any of the *Koelreuteria* add a dramatic splash of color to the late season landscape, when few other trees or shrubs are flowering. They can be planted in masses for a very dramatic effect at bloom time or planted singly in a lawn or by a patio where it makes a lovely specimen tree. Either of the golden-raintree species' loose clusters of rich golden blooms atop light summery foliage give way to ornamental fruits and clear yellow fall color. Plant these beautiful trees and let their bright rain of blooms shower your landscape with late summer gold.

Lagerstroemia fauriei / Japanese crape myrtle   PLATE 87

Just when it seems we must turn to herbs and flowers for summer color, the bright, ruffled blooms of crape myrtle burst into pinks, lavenders, and white all over southern gardens, parks and streets. Traditionally, crape myrtle plantings have been made up of the common crape myrtle, *Lagerstroemia indica*, which is native to China and Korea. But Japanese crape myrtle, *Lagerstroemia fauriei*, offers other interesting attributes.

Lagerstroemia fauriei is a larger, more cold-hardy species than common crape myrtle, and it shows off its frilly, white blooms in early summer, before other species. This multistemmed small tree reaches 20–30 feet (6.2–9 m) in height with an upright, arching habit (with some variability among seedling trees). The neat, light green foliage turns a pleasant canary yellow in the fall. The star attraction, however, is the exfoliating bark. As the plant matures, the smooth, older gray bark peels away to reveal large swathes of bright cinnamon new bark—a breathtaking effect.

Lagerstroemia fauriei is native to Japan, where it was first collected by John Creech in 1956 in the mountains. Because it is native to a colder climate than *L. indica*, Japanese crape myrtle is hardier. In fact, *L. fauriei* survived the severe winters of the mid-1980s, which killed many *L. indica* plants to the ground.

Crape myrtles, in general, thrive in sun and heat, provided there is adequate water. Powdery mildew can be a problem on *Lagerstroemia indica* plants, but an added bonus

PLATE 52. Hercules' club (*Aralia spinosa*) is a ough, durable plant easily used to create un-sual texture.

PLATE 53. The bold habit, flowers, and variegated foliage of *Aralia elata* 'Variegata' make a strong statement in the landscape.

PLATE 54. *Buddleia globosa* is the only orange-flowered butterfly bush with rounded flower clusters.

PLATE 55. The orange-sherbet colored flowers of *Buddleia globosa* contrast pleasingly with its grass-green foliage.

PLATE 56. The hybrid 'Madame Galen' combines the showy flower display of *Campsis grandiflora* with the deep orange-red color of *Campsis radicans*.

PLATE 57. The foliage and fruit of *Carpinus betulus* 'Pendula' have an interesting texture.

PLATE 58. *Carpinus betulus* 'Pendula' makes a magnificent specimen or lawn tree.

PLATE 59. *Caryopteris* 'Worcester Gold' has unusual lime-gold foliage that can be massed to create a bright glow in the garden.

PLATE 60. The weeping *Cedrus deodara* 'Pendula' is an especially graceful conifer.

PLATE 61. *Cercidiphyllum japonicum* 'Pendula' is an unusual weeping form of katsura tree.

PLATE 62. The flowers are only one of the attributes of the indestructible *Clerodendron trichotomum*.

PLATE 63. The metallic blue fruits of harlequin glory-bower are especially unusual. (KIM TRIPP)

PLATE 64. 'Hummingbird' is a particularly floriferous, compact growing form of *Clethra alnifolia*.

PLATE 65. *Clethra alnifolia* 'Pink Spires' has light pink flowers.

PLATE 66. *Cotinus coggygria* 'Daydream' has unusually dense and compact inflorescences. (KIM TRIPP)

PLATE 67. *Cunninghamia lanceolata* 'Glauca' is the blue foliaged form of Chinafir.

PLATE 68. *Cyrilla racemiflora* offers graceful flowers and foliage in summer.

PLATE 69. The infructescences of *Euscaphis japonica* are brilliantly colored.

PLATE 70. The fruit clusters of sweetheart tree contrast brightly with the rich green foliage to make a showy display.

PLATE 71. The striped bark of sweetheart tree is ornamental in all seasons of the year.

PLATE 72. The exotic blooms of *Feijoa sellowiana* add tropical character to the garden.

PLATE 73. The drying seed capsules of Chinese parasol tree are the inspiration for its common name.

PLATE 74. *Gordonia lasianthus* has a refreshingly clear white flower even in summer.

PLATE 75. This hedge of adult English ivy at Kew gardens is completely self-supporting and stands alone on its own wood with no wall or poles for support.

PLATE 76. The rose-of-Sharon cultivar 'Blue Bird' is the closest to a true-blue flower.

PLATE 77. 'Unique' is an excellent selection of *Hydrangea paniculata* with an extended period of bloom.

PLATE 78. Oakleaf hydrangea (*Hydrangea quercifolia*) in flower is informal but eye-catching.

PLATE 80. *Hypericum frondosum*, even before peak bloom, is a lovely low shrub. (CLAIRE SAWYERS)

PLATE 79. Oakleaf hydrangea has excellent fall color as well as a showy flowering display.

PLATE 82. The weeping yaupon holly (*Ilex vomitoria*) displays its bright red fruit on downwardly arching branches.

PLATE 81. The flowers of *Hypericum frondosum* have a dense, thick cushion of yellow stamens in the center of the flower. (CLAIRE SAWYERS)

PLATE 83. The rose-purple flowers and finely divided foliage of *Indigofera* are a welcome addition to the late summer garden.

PLATE 84. 'Henry's Garnet' sweetspire (*Itea virginica*) not only flowers profusely but has burgundy-red fall color.

PLATE 85. 'Fukurin' variegated kadsura vine (*Kadsura japonica*) keeps its distinctive foliage quality all summer.

ATE 86. The seedpods of *Koelreuteria bipinnata* mature to a soft
lmon pink.

PLATE 87. 'Fantasy' Japanese crape myrtle (*Lagerstroemia fauriei*) combines the showy exfoliating cinnamon bark of the species with a distinctively narrow vase-shaped habit.

PLATE 88. Shrub bush clover creates fountains of purple flowers when it blooms in summer. (KIM TRIPP)

PLATE 89. 'Little Gem' is a unique form of *Magnolia grandiflora* that blooms at a young age.

PLATE 90. Southern magnolia is one of the ever-green 'grande dames' of warm climate gardens.

PLATE 91. Virginia pine is a beautiful small native pine that thriv‹ in many soils and climates.

PLATE 92. The semi-flattened needles of Virginia pine are relatively fine textured and gardener-friendly.

PLATE 93. The long chains of ripening fruits Chinese wingnut (*Pterocarya stenoptera*) are e‹ pecially showy.

PLATE 94. Chinese wingnut is a fast-growing, exceptionally tough shade tree.

PLATE 95. White oak (*Quercus alba*) is famous for its light, silvery bark and spreading crown.

PLATE 96. The rounded crown and narrow leaves of willow oak (*Quercus phellos*) make it a handsome shade tree with broader utility than is currently recognized.

PLATE 97. Fastigate English oak (*Quercus robur* 'Fastigiata') has a formal, upright habit.

PLATE 98. The evergreen foliage of Virginia live oak stays a deep, glossy green all year.

PLATE 99. _Rhododendron prunifolium_ bears carmine flowers in late summer when few other shrubs are in bloom.

PLATE 100. 'Moonlight' is a lovely selection of _Schizophragma hydrangeoides_ selected for its silvery blue-green foliage. (KIM TRIPP)

PLATE 101. Climbing hydrangea, _Hydrangea anomala_ ssp. _petiolaris_ makes a magical entranceway.

PLATE 102. Pagoda tree (*Sophora japonica*) in flower is a lovely, refined shade tree.

PLATE 103. The lavender-pink flowers and blue-green feathery foliage of tamarisk (*Tamarix ramosissima*) make it a peacock in the garden.

PLATE 104. The native bald cypress (*Taxodium distichum*) makes a majestic specimen in summer and winter.

PLATE 105. The deciduous foliage of *Taxodium distichum* is fine textured and medium green in summer but turns bright cinnamon in the fall before it drops.

PLATE 106. Lindens (*Tilia* spp.) are staples of the modern landscape, but we rarely notice the ornamental qualities of their flowers and fruit.

PLATE 107. 'Lace Parasol' is a magnificent weeping selection of the native *Ulmus alata*, distributed by The NCSU Arboretum.

PLATE 108. Both white and purple selections of *Vitex* are easy to use in the landscape.

PLATE 109. The lavender flower spikes of *Vitex* combine readily with many perennials.

of Japanese crape myrtle is its resistance to this pervasive fungus. To maximize the mildew resistance of _L. fauriei_, plant this tree where there is plenty of air movement. _Lagerstroemia fauriei_ is a relatively foolproof tree which performs well, given full sun. It can be limbed up even as a young tree to maximize the view of the smooth, undulating trunks and their dramatic bark. Japanese crape myrtle is cold-hardy to zone 6—decidedly more cold-hardy than the zone 7 hardiness of _L. indica_. Propagation of _Lagerstroemia_ is successful from seed, which is held at 40°F (4°C) for a month, or from summer softwood cuttings, which are treated with rooting promoters and rooted under mist. _Lagerstroemia fauriei_ tends to be more difficult to root from cuttings than _L. indica_.

Lagerstroemia fauriei_ has been an important part of the crape myrtle breeding program at the United States National Arboretum, out of which many _L. fauriei_ × _L. indica_ hybrid cultivars have come. This group of widely available hybrids, including the white-flowering 'Natchez' and the lavender 'Muskogee', combine the beautiful bark and the mildew resistance of _L. fauriei_ with the added range of flower color of _L. indica_.

Some of the seedlings from famous plantsman John Creech's original collection of _L. fauriei_ from Japan have been in evaluation at the North Carolina State University Arboretum. 'Townhouse' is a seedling with dark wine-red bark and small, compact stature—a fabulous tree for a townhouse garden. Another beautiful cultivar, named 'Fantasy' (Plate 87), has an unusual vaselike habit, reminiscent of American elm (_Ulmus americana_). This pure _Lagerstroemia fauriei_ cultivar has the incredibly beautiful bark, lovely white flowers, and exceptional hardiness and vigor of the original collections. One of the most memorable garden experiences I have ever had was to come around a bend in the path at the North Carolina State University Arboretum to catch the rays of the setting sun as they lit up the trunk of 'Fantasy' Japanese crape myrtle. The cinnamon bark glowed orange and crimson as if on fire, illustrating what a truly stunning tree Japanese crape myrtle is for the summer landscape.

Lespedeza bicolor / Shrub bush clover   PLATE 88

The sprightly blooms of _Lespedeza bicolor_, shrub bush clover, mark the turning of the seasons from summer toward fall with a fountain of rosy pink and lavender. Shrub bush clover is native to North China and Japan but has naturalized in the United States since its introduction in the mid- to late 1800s. In addition to its use as an ornamental, it was planted in many areas for erosion control. As a result, naturalized plants can be found growing along roadsides, in fields and clearcuts, and in any open, well-drained sites.

As its name suggests, shrub bush clover is related to common clover, but it looks very different from the lucky little plants of field and meadow. Like its common clover cousin, shrub bush clover has a trifoliate leaf, but that's where the family resemblance ends. _Lespedeza bicolor_ is a deciduous shrub that grows 6–9 feet (1.8–2.7 m) in height. The plant has a very open and spreading form. The profusion of late summer flowers sprinkled throughout its dark green, loose-textured foliage are the primary attraction of shrub bush clover. The flowers, which resemble those of the garden pea, are a lovely blend of pink with soft purple in the center.

Shrub bush clover is very easy to cultivate and grows rapidly in any sunny, well-drained site. It has no significant disease or insect problems and is cold-hardy to zone 4. It is best used as a herbaceous perennial, or die-back shrub, in colder areas of zone 4. Cut back to about 18 inches (45.6 cm) off the ground and mulch in late fall. One of the most appealing aspects of this plant is its loosely spreading and informal shape, but the plant can be pruned to keep it from becoming too untidy in smaller spaces. It is readily propagated from directly sown seed or softwood cuttings.

Two cultivars are commercially available. 'Summer Beauty' has an extra-long flowering period, from July through September in zones 5–8. 'Yakushima' is a dwarf that grows in a tight, 1-foot-tall (0.3-m) mound with smaller, though equally attractive, flowers and foliage than the species. There is also a second excellent species for the garden, *Lespedeza thunbergii*, which is similar to *L. bicolor* in many ways. It differs in having more refined, bluish leaves and darker rose-purple flowers usually borne in greater profusion.

Shrub bush clover is an unlovely name that does not do justice to its owner. The lazily arching branches, freckled with bunches of rosy-lavender flowers, offer a delight in the garden that reflects our own softened mood at the end of summer. This is one of the most beautiful and useful informal shrubs for massing. It is dense enough to make an effective screen, yet the fine texture of the foliage and lightness of the flowers keeps it from feeling heavy. It is wonderful in a mixed border or as an anchor for a perennial border. It makes a fine seasonal screen, or can be used as a tall groundcover on banks and slopes or as a cascade of rosy color near a porch.

Magnolia grandiflora / Southern magnolia   PLATES 89, 90

The quintessential southern landscape rolls between lush woods and red clay fields and is punctuated by grand old southern magnolias spreading their evergreen skirts all the way to the ground—the envy of Scarlett O'Hara herself! *Magnolia grandiflora* is a tree of such distinction and beauty that it has become a veritable icon of the grace and charm of the South, both old and new.

This distinguished southeastern native is well known as a tall and imposing broadleaf evergreen tree, reaching heights of 80 feet (24.8 m) with age, and developing a rounded, conical shape. Southern magnolia is also famous for its incredibly wide, fragrant, creamy flowers that open to initiate the beginning of every summer, spreading a heady aroma on the warm, humid air. These blooms mature into cone-shaped spirals of bright scarlet berries that adorn the trees through the fall and into winter. The beautiful leaves, often as large as 12 inches (30.4 cm) long and half as wide, are dark moss-green and shiny on the upper surface. The lower surface of the foliage is usually covered with a cinnamon-brown pubescence, a fuzzy layer sometimes called the "ruff" or "indumentum."

Southern magnolia is generally cold-hardy to zone 7 with some forms being reliable in zone 6 ('Edith Bogue' and 'Victoria' are especially hardy selections discussed below that make good choices for colder areas). It is well adapted to zones 7 and warmer in areas that receive summer and winter rainfall. It is very tolerant of shady areas but will not flower as profusely or have as dense a habit in the shade. There are no significant disease or pest problems with this sturdy tree. Large trees may drop significant numbers of leaves after transplanting, so it is best to move and plant young trees.

For over a century, *Magnolia grandiflora* was generally propagated by seed, and the majority of trees planted were from quite variable populations of seedlings. Many plantings exhibit a broad range of heights, shapes, foliar color, and earliness and profusion of flowering (seedling southern magnolias can take upwards of a decade to begin flowering, or they may start as early as the tender age of two to three years).

There are, however, many cultivars of *Magnolia grandiflora* with consistent character that have been neglected in production because of the difficulty of propagating these selections. This situation has changed as more and more nurseries become expert at grafting these various cultivars and as techniques have been developed for successfully rooting cuttings.

A number of good cultivars of southern magnolia that deserve far greater attention and use than they have received in the past are now available for the landscape. 'Bracken's Brown Beauty' is a bit of a tongue-twister for a name, but the compact growth habit combined with smaller but full foliage makes this a very attractive plant. The leaves often exhibit slightly wavy margins. The name honors the founder, Ray Bracken of South Carolina, and refers to the heavy orange-brown pubescence on the undersides of the leaves. 'Claudia Wannamaker' has very dark green, lustrous foliage with warm brown undersides. Growth habit is more open and less formal than other southern magnolias. This cultivar flowers reliably early in development. 'D. D. Blancher' is a relatively upright form with good-quality foliage and warm brown leaf pubescence. 'Edith Bogue' is a pyramidal selection with unusually narrow leaves and exceptional cold hardiness; it thrives as far north as Philadelphia. 'Glen St. Mary' (or 'St. Mary') is an early-flowering form with bronzy undersides to the leaves. It has been advertised as compact, but this form has grown at rates close to those of most other _Magnolia grandiflora_ selections at the North Carolina State University Arboretum in zone 7. 'Gloriosa' has very broad foliage and probably the largest flowers, reaching up to 12 inches (30.4 cm) in diameter with thick, waxy petals. 'Goliath' is another large-flowered form with very little pubescence on the leaves. 'Hasse' is particularly interesting because of its upright growth habit and small, neat, very glossy leaves. 'Lanceolata' (sometimes seen as 'Exmouth' or 'Exionensis') is probably the oldest known cultivar, having been in cultivation for 200 years. The foliage is quite narrow and lance-like. The flowers of 'Lanceolata' appear semidouble because of another set of "tepals" (magnolias' primitive form of petal) that ring the center of the bloom.

'Little Gem' (Plate 89) is a very exciting cultivar, selected by Warren Steed of Steed's Nursery in Candor, North Carolina. This is a smaller-leaved, relatively compact form, an excellent choice for today's smaller landscapes. It is sometimes referred to as a dwarf cultivar, but this is not the case—it has reached 12 feet (3.7 m) in six years at the North Carolina State University Arboretum in Raleigh, North Carolina. Nonetheless, the relatively compact form of 'Little Gem' can bring the aristocratic character of southern magnolia to gardens without the great need for space required by the species or full-sized cultivars. 'Little Gem' has the added advantage of flowering as a young tree. Its extended bloom time lasts throughout the summer and fall, although there is some lessening of bloom during the summer's peak of heat. It is important to note that 'Little Gem' is one of the least cold-hardy of these selections and will not thrive further north or west than zone 7.

The cultivar 'Majestic Beauty' combines huge leaves and very profuse flowering with a tall, pyramidal form. 'Ruff' intrigues with reddish pubescence on the foliage. 'Russett' is a cold-hardy selection with small, very dark leaves. 'Samuel Sommer' is relatively rapid in growth, with large flowers, and eventually becomes a large, handsome tree. 'Symmes Select' is a Cedar Lane Farm selection from Georgia with full branching all the way to the ground, and large flowers even on young trees. 'Victoria' is a medium-sized cultivar from Vancouver, British Columbia. Reported to be among the hardiest of all southern magnolia cultivars, 'Victoria' is very popular in the Pacific Northwest.

Few other trees can ever reproduce the stately grandeur of a mature southern magnolia. Yet the magnificent _Magnolia grandiflora_ is deserving of far greater attention in modern landscapes than it currently receives. The compromises modern gardeners require, of predictable growth habit and more compact size, can be met with the named selections. By planting these cultivars of southern magnolia, the gardener continues the tradition of planting for a future of grace and charm.

Pinus virginiana / Virginia pine   PLATES 91, 92

Pinus virginiana is a small tree suited to modern landscapes that combines reliability on clay or poor soils with sculpted landscape character. Young trees are broadly pyramidal with somewhat open branching. As the tree ages, the crown becomes flattened and umbrella-like. Unlike many pines, this species often retains its lower branches throughout the life of the tree, and they become gnarled and twisted with age. As the canopy becomes flattened, these lower branches become visible and add sculpted interest to the landscape.

Virginia pine is naturalized throughout the mid-Atlantic and southeastern regions of the United States. It is well adapted to a range of conditions, including heavy, wet, clay soils to sandy loams. It is a small tree by pine standards, reaching 20–40 feet (6.2–12.4 m) in height with a canopy spread of up to 30 feet (9 m) on older specimens. Bark is a smooth, light gray on young trees, and it eventually darkens and forms scales with red-brown fissures as trees mature. The foliage is a glossy, deep green that lightens in color during the winter. Needles are relatively short, about 2 inches (5 cm) long, and carried in bundles of two (Plate 92). The needles of Virginia pine are slightly twisted so that each needle looks as if a gardener had pinched the tip and twirled it in a thoughtful moment. The dark brown cones are borne on two-year-old wood and often persist for several years after they open.

Fully hardy to zone 4, Virginia pine is very tolerant of difficult sites. It grows slowly but will perform well in areas where other conifers cannot grow. This beautiful tree can thrive in conditions ranging from poor, heavy soils to windswept hillsides. Its one requirement is an open area in full sun. This is an excellent tree for new, bare landscapes, because it will tolerate the stressful conditions often associated with the scraped soils of new landscapes and will thrive in the full sun of an empty site. In addition to use as a single tree, Virginia pine makes effective plantings in small groups of two or three. It can also be massed in greater numbers on larger properties, where the trees help create sheltered areas for less stress-tolerant plants.

Pines are generally difficult to propagate vegetatively and are commercially grown from seed or grafted. Virginia pine germinates readily from seed, but if propagation of a particular character is desired, scion wood of the selected plant is grafted onto seedling rootstock. Named cultivars of Virginia pine are often difficult to obtain. 'Hess's Weeping' is a prostrate form, which grows along the ground with pendulous branching. The spectacular 'Wate's Golden' has brilliant gold foliage in winter, turning back to green in spring and summer.

Virginia pine has been called scrub pine and is sometimes considered a poor cousin to other pines available for use in the landscape. This is an undeserved reputation for an evergreen conifer with a versatile and adaptive nature. Dependable *Pinus virginiana* is ideal for today's restricted, demanding landscapes. Near the entrance to the western half of the North Carolina State University Arboretum, Virginia pine has demonstrated its versatile and adaptable nature—just the kind of character we need in the garden.

Pterocarya stenoptera / Chinese wingnut   PLATES 93, 94

The canopy of shade trees is welcome in summer, even late in the day, to cool the heat of the sun's rays and keep the garden a pleasant place to be. Choosing, planting, and establishing shade trees requires thought, creativity, and an abundance of hard work. _Pterocarya stenoptera_, Chinese wingnut, is an excellent shade tree for difficult new landscapes that lack large trees. It is an extremely tough, rapid grower that will fill a barren space quickly with soft, beautiful growth of up to 6–10 feet (1.8–3.1 m) a year. In an amazingly short timespan, Chinese wingnut can add important structural elements to a new landscape that only large shade trees can provide. It often reaches very large dimensions in the landscape and is therefore not a good choice for townhouse, condominium, or other small gardens in confined spaces. It is just this rapid growth, however, that makes it such a good choice for new, bare areas, for quick or temporary replacement of trees lost to unpredictable weather or disease, or for urban parks and other large, difficult green spaces.

Introduced from China in the late 1800s, Chinese wingnut is a relative of walnuts and pecans, and it carries a certain family resemblance to its cousins. It is a tall, spreading, sometimes multitrunked tree reaching 50–70 feet (15.2–21.7 m) in height and spreading to 40 feet (12.4 m). The tree has open, somewhat pendulous branching and the walnutlike compound leaves are a soft, lustrous grass-green. The open canopy and foliage cast dappled shade. As the tree grows, the expansive roots may emerge from the ground in the manner of old willow oaks (_Quercus phellos_). The rough bark is an attractive gray-brown. One of the most beautiful features of this durable tree is its fruit. Throughout the summer, long necklaces of emerald-green, winged nutlets are draped from its drooping limbs and often hang all the way to the ground. These seemingly fragile chains, first of tiny flowers in the spring, then of delightful fruits in the summer, decorate the tree until fall and may be up to 2 feet (0.6 m) long. The fruits turn from emerald to golden brown before they fall in September.

Pterocarya stenoptera is hardy to zone 6. It grows fastest in moist, rich, well-drained loam, but it will also grow rapidly in dry, exposed sites and is tolerant of a wide range of pH and soil conditions. It does best in full sun. It is important when transplanting to allow a wide enough space for the extensive root system. Successful propagation by seed requires a two- to three-month cold treatment before germination. For vegetative propagation, summer softwood cuttings are treated with concentrated rooting promoter and rooted under mist.

Three other _Pterocarya_ species are available. _Pterocarya fraxinifolia_, Caucasian wingnut, is very similar to Chinese wingnut and may easily be confused with it. Specimens sold as _P. fraxinifolia_ may indeed be _P. stenoptera_. The winged nutlets of Chinese wingnut have longer wings, while those of the Caucasian wingnut are more horned in appearance. A hybrid between these two wingnuts is _Pterocarya × rehderiana_ which has improved hardiness and even more vigor than its parents. _Pterocarya rhoifolia_, the Japanese wingnut, has wingless nutlets and extremely long leaves.

Chinese wingnut can rapidly turn a barren area into a welcoming green space. This tough tree can be used in the most challenging of landscapes, parks, and ambitious gardens. It is the perfect shade tree to help make our urban deserts into oases of parks and gardens.

Quercus alba / White oak   PLATE 95

The magnificent crowns of *Quercus alba*, white oak, spread themselves over fields and drives throughout the eastern and central United States and around the Great Lakes in Canada. Many of these ancient doyennes are upwards of 100 years old, reaching 60–80 feet (18.6–24.8 m) in height with spreads of up to 100 feet (31 m).

The spreading profile of white oak has been used as the quintessential tree of art, sign, and symbol for hundreds of years. Two round-lobed white oak leaves crossed over its classic acorn have also frequently been used to indicate forests and nature in general.

The oaks are a generally confusing group, but the light gray, platy bark of white oak and its distinctive round-lobed leaves make it relatively easy to separate this species from its cousins. The glossy, deep green leaves remain dark green, with bluish overtones, all through the summer, changing to a deep wine or purple-red in the fall.

Quercus alba takes years to reach the great size of maturity, but it grows quite rapidly as a young tree and at any age is one of the most beautiful trees found in the landscape. Hardy to zone 3, it is very well adapted to a range of soils and conditions. It can thrive in both clay soils and loamy sands and, once established, will tolerate drought; however, it does not tolerate having its roots disturbed once planted. That is why fine old specimens are scattered throughout rural areas and only in urban sites that have not seen new construction.

Why do we not see white oak grown everywhere that a classically beautiful large tree is needed? Why isn't white oak planted and maintained in new landscapes, parks, cemeteries, and golf-courses? One reason is because white oak is propagated by seed. Because it takes longer than other plants to reach a saleable size, white oak is not a profitable proposition for nurseries and garden centers. White oak also has a somewhat exaggerated reputation for being difficult to transplant and move. While it is true that large, field-grown trees do not take kindly to being dug and moved, transplanting is not a problem if young trees are moved from containers or balled-and-burlapped.

Another reason white oak has been neglected is because it has a reputation for being "slow" and "difficult" in the landscape. This is not accurate. As a young tree, white oak grows about as rapidly as the fast-growing willow oak (*Quercus phellos*), but it takes many years to reach the majestic size of older specimens because growth rate slows after the first decade. A large part of the reason for white oak's reputation as difficult is because existing white oaks are usually abused to death in new construction sites. Venerable trees are fenced close to their trunks, and massive equipment is run over and over and around the roots in all kinds of weather. Any remaining organic matter and topsoil is scraped away, taking many of the tree's feeder roots with it, and is sold back to the new buyer or hauled to a different site. In this fashion, what may have been previously an established, rich clay soil with good fertility, pH, moisture conditions, respectable organic matter content (from many years of leaf fall under the white oak), full of oak roots, becomes transformed into a root-poor, bricklike desert of compacted soil that floods with the first heavy rain. White oaks already on the property spend the next five years dying, accompanied by a host of secondary pests and diseases, which are blamed on the tree's "troublesome" nature. In contrast, young trees planted and cared for in new landscapes are tough, beautiful investments.

Quercus alba is a low maintenance tree. It does not require extensive pruning. The wood of oaks is very hard and strong, so long, thick, outspreading limbs are dramatic, not dangerous. The stout branches are excellent candidates for swings and shady picnic spots. It is extraordinarily rare for a white oak limb to come down in an ice or wind storm. A large white oak near the home is a blessing, not a liability.

It is a revealing commentary on our modern perspective that we rarely plant or preserve large trees with an eye toward the landscapes of our children. White oak is one of the most magnificent trees of North America. It is a living reminder of the reverence for landscapes, both of the past and the future, that we will need if we are to keep our living spaces as gardens for today and tomorrow.

Quercus phellos / Willow oak   PLATE 96

From porches to parking lots, shade is at a premium in the heat of summer. Suddenly shade trees seem much more appealing than they did in January, and we wonder why we never planted glades of them in every open site. One reliable shade tree is the underappreciated and excellent willow oak, *Quercus phellos*. This native of the eastern United States has long, narrow leaves that, from a distance, appear nearly identical to those of weeping-type willows (e.g., *Salix alba*).

Willow oak is unique among oaks because of its fine-textured foliage and its relatively rapid growth rate. Like other oaks, it is a very large tree; unlike other oaks, it reaches mature size rather quickly and is therefore not a good choice for small sites. These qualities, however, make willow oak an excellent street tree or a perfect choice for properties where large, trouble-free trees are desired.

Quercus phellos will ultimately reach 50–70 feet (15.2–21.7 m) in height with a nearly equal spread. Growth rates approach 2 feet (0.6 m) a year on young trees in suitable sites. With some judicious pruning, willow oak can develop the classic spreading canopy found in its close relatives, such as white oak (*Q. alba*); otherwise, it develops a more rounded canopy. As the tree matures, it develops a stout trunk covered with finely grooved, light charcoal-gray bark. The fine-textured leaves are bright grass-green in color, and the foliage retains good quality throughout the summer. Fall color of sandy yellow to russet tan is appealing but not as spectacular as that of some other oaks.

Willow oak is generally prolific in acorn production, often with alternate years of heavy and light crops. Its acorns are small and fine in appearance, like the tree's foliage. Their startling bright orange flesh always seems to fascinate children who happen upon crushed acorns on walks and drives.

Quercus phellos prefers moist, well-drained soils in full sun, but it is amazingly resilient and adaptable and will thrive in essentially every landscape site, now matter how tough or urban, given adequate light. It has no significant pest or disease problems and is tolerant of pollution and drought. It is completely hardy to zone 5. Willow oak transplants more easily than some other oaks as it has a more fibrous, rapidly growing root system (this is important to consider when siting trees near sidewalks and buildings to avoid root eruptions through shallow paving, such as sidewalks or patios). *Quercus phellos* is readily propagated from seed. Fresh seed is sown directly outside, to germinate in spring, or seed is harvested, stored in moist medium at 40°F (4°C) for one to two months, and then sown in pots. Willow oak is one of the easiest oaks to vegetatively propagate. Summer hardwood cuttings from young trees are rooted under mist.

Quercus phellos has been used historically as lawn, street and park tree. Many of these plantings have matured into outstanding allées of fine-textured large trees. In some cities, unfortunately, the super-adaptive nature of willow oak seems to have led us to take this tree for granted, and older plantings are occasionally cut down without appropriate consideration. In the right place, this indestructible tree is a treasure.

Quercus robur 'Fastigiata' / Columnar English oak   PLATE 97

"The mighty oak" brings to mind an ancient monarch of trees, spreading broad and stately branches in all directions, its massive crown atop a thick and imposing trunk. Many of the world's roughly 275 oaks do indeed grow to achieve such venerable dimensions, but there are exceptions that can be very important for different settings. A unique upright form of the English oak, *Quercus robur* 'Fastigiata', offers the qualities of an English oak but requires less landscape area than its broad-spreading parent tree. While it is not a small tree, eventually attaining 40–80 feet (12.4–24.8 m) in height, its growth is mostly vertical, with a spread of only about 10 feet (3.1 m). This special habit makes the columnar English oak an excellent choice for modern landscapes, where there is often far more vertical space than ground area.

The shape of columnar English oak is upright, but not excessively narrow. The foliage is typical of English oak. Leaves are deep green, almost bluish, in color with the classic oak leaf shape softened by very rounded lobes. The summer foliage is among the most beautiful of the deciduous oaks. The dried, light brown leaves usually persist on the tree all winter and are shed in the spring as the buds expand into new growth. While fall color is not of interest, the glossy brown acorns are attractive. Usually a full inch (1.2 cm) in length, the nuts themselves are much longer and showier than the acorns of many other oak species.

Columnar English oak is hardy to zone 4. It prefers well-drained soils but will tolerate heavy clay conditions as well. Many oaks require decidedly acid conditions to prevent chlorosis, but *Quercus robur* 'Fastigiata' is tolerant of a significantly wider range of soil pH. Full sun is important for healthy growth and foliage quality. Columnar English oak is susceptible to powdery mildew late in the summer, as the fall approaches, and should be grown in areas with maximum air flow. Although commonly found on the tree, the powdery mildew causes only cosmetic damage to the plant.

Columnar English oak can be propagated readily by various methods. Unlike many horticultural selections, it will come relatively true from seed (and is therefore sometimes referred to as the botanical variety, *Quercus robur* f. *fastigiata*, instead of a horticultural cultivar). Acorns are easily collected in the fall and germinate readily, as they have no dormancy or other preconditioning requirements. Seedlings will exhibit a certain range in the degree of the fastigiate character, but 80–90% of the seedlings should be definitely fastigiate. Undesirable plants are removed from the seedling population. Columnar English oak can also be grafted onto rootstock of the species *Q. robur*. Summer softwood cuttings can be rooted under mist after treatment with concentrated rooting promoter; because of low success rates, this is not a preferred method of propagation.

Columnar English oak is an outstanding large tree that combines special beauty with a useful habit for modern gardens. In larger areas, the trees can also serve as a large hedge or screen, if set 8–9 feet (2.4–2.7 m) apart in a row. Custom-sized for today's gardens, the vertical form of columnar English oak adds vertical dimension, even in landscapes where space is limited.

Quercus virginiana / Live oak   PLATE 98

Live oak is the quintessential tree of southern estates. Lining a curving drive, draped with gray Spanish moss, this statuesque tree recalls a bygone era in architecture and landscape alike. *Quercus virginiana* is an evergreen oak, native to the Deep South of the United States and west into Mexico. It gets very large with age, developing curving, sculpted branches that intertwine and arch over into neighboring trees, creating the famous oak allées of southern plantations.

Live oak can reach heights of 50–80 feet (15.2–24.8 m), with a seemingly impossible spread of over 100 feet (31 m) in width. The neat, relatively small leaves are a very glossy, dark black-green. The foliage remains dark green on the tree all year long, adding wonderful winter character. Acorns of this oak are as demure as the foliage, being only .75 inch (1.8 cm) long. The very dark, nearly black nut is almost covered by its rough cup. An unusual character of the acorns is that they are borne at the end of a 1–3-inch (2.5–7.6-cm) stalk. The dramatic bark of live oak is a very dark charcoal color. As the tree matures, the bark develops a furrowed, chunky appearance.

Live oak is often thought of as reliably hardy only to zone 7 or 8. This is not necessarily the case. Hardiness depends on the area of origin, or provenance, of a given seedling. If a seedling tree is from one of the northern populations of live oak, from Virginia, for example, the tree can be completely hardy through zone 7. On the other hand, if the seedling is from a more southerly population, from southern Louisiana, for example, it will not be reliable hardy outside of zone 8. The live oak trees at the North Carolina State University Arboretum in Raleigh, North Carolina, from relatively northern races, survived the extreme freezes of the mid-1980s and remained beautiful, while other trees in Raleigh, from Deep South races, were killed by those same freezes.

Live oak prefers moist soils in full sun or partial shade but will tolerate almost any soil from light sand to very heavy clay. It should be planted as a young tree to avoid serious transplant shock. _Quercus virginiana_ is usually propagated from seed, but it can also be grown from semihardwood cuttings, unlike most other oaks. Cuttings taken from young trees in late spring, summer, and early fall are treated with relatively high concentrations of rooting hormone and rooted under mist.

No cultivars of live oak are in commercial production, but there are botanical varieties whose populations overlap with the species in the wild. _Quercus virginiana_ var. _fusiformis_ is a smaller, multistemmed form found growing in Texas. _Quercus virginiana_ var. _maritima_ grows along the coast of the Southeast on barrier islands and right near the shore.

Live oak is not a tree for small properties, but it deserves wider use in large-scale settings. Few trees can match its character. It makes an unforgettable street tree, and is surprisingly tolerant of many stressful conditions, including salt spray.

Rhododendron prunifolium / Plumleaf azalea   PLATE 99

Flowering trees and shrubs are scarce in late summer, and even the perennial palette is strained after the early season peak and before the asters and goldenrods of fall begin blooming. Certainly the rhododendrons and azaleas are long since finished . . . but wait! What is that flash of scarlet in the woods, along the bank, by the creek? Surely not a native azalea at this time of year?

Your eyes have not deceived you: the brilliant blossoms belong to the native azalea _Rhododendron prunifolium_, plumleaf azalea, which saves its display until late summer. This deciduous shrub reaches 6–12 feet (1.8–3.7 m) in height with medium texture and unassuming foliage. Otherwise modest, plumleaf azalea comes into its own in late summer, when 2-inch (5 cm) floral trumpets create bursts of scarlet scattered throughout the branches. The display is a welcome bright splash in the woods and along streambanks of much of the eastern United States.

Plumleaf azalea is hardy to zone 5, but it may suffer some damage in the most exposed sites and is not a good choice for seaside gardens. It thrives in a range of soils, from light sand to all but the worst clay. Partial shade gives best results in warmer regions, but too much shade will result in few flowers. The plant is propagated from seed or from softwood cuttings rooted under mist.

Several cultivars of plumleaf azalea have been selected for degree of showiness and color variations. 'Cherry-Bomb' certainly has the name with the greatest fun factor; its blossom are vivid orange red. 'Coral Glow' has pink-orange blooms with a salmon tint. 'Lewis Shortt', named for its founder, offers true red flowers and foliage that looks nearly identical to the leaves of plum trees. 'Peach Glow' blooms are peachy orange in color. 'Pine Prunifolium' has flowers of a clear, bright red.

The unique display of bright flowers, as well its surprisingly tough performance, makes *Rhododendron prunifolium* a valuable addition to the summer garden. Its vivid blossoms create a flash of visual energy for visitors in the garden, transforming the doldrums of summer with their unexpected vibrancy. Plumleaf azalea is an excellent choice to bring summer color to suburban lawns or the edge of woods and hedges. It is most effective used in an informal mass or scattered under trees, and it combines nicely with gold-flowered summer perennials, such as *Rudbeckia*, in a hot border.

Schizophragma spp.; *Hydrangea anomala* / Climbing hydrangea

PLATES 100, 101

Woody vines make a graceful contribution to summer's cloak of green, but they are often overlooked in the splash of color from flowering trees, shrubs, perennials, and bulbs. Yet, as the season progresses, the flowing outlines and interesting flowers of many vines can provide a special feeling in the garden. Some of the most lovely vines are a group of plants called the climbing hydrangeas.

Climbing "hydrangeas" include several different but related plants, native to China and Japan, with similar but varied habit, form, and foliage. Two true hydrangeas are included in this group, the climbing hydrangea, *Hydrangea anomala* ssp. *petiolaris*, and the species itself, *Hydrangea anomala*. Two other climbing "hydrangeas" are actually not hydrangeas at all, but are members of the related genus *Schizophragma*.

Climbing hydrangea, *Hydrangea anomala* ssp. *petiolaris* (Plate 101), is a vine that can climb almost anything, from a rock or brick wall to a telephone pole. It can reach 80 feet (24.8 m), given a structure to hold it. The lovely white flowers are borne in flat-topped clusters of showy outer flowers and inconspicuous inner flowers, much like many of the viburnums. The flower clusters are carried nearly horizontal, scattered at different levels up the length of the vine in many layers, which creates a three-dimensional effect. The rich green foliage is also carried somewhat horizontally and in various planes, adding to the spatial interest of the vine. The leaves are rounded-oval in shape and coarsely toothed. The foliage remains handsome throughout the growing season and maintains its quality right up until leaf fall in the autumn. The bark on older plants develops a peeling habit and is a beautiful cinnamon color, which adds to winter interest.

Hydrangea anomala is different from the climbing hydrangea, *H. anomala* subsp. *petiolaris*, in that its leaves are narrower and more oval in shape, and the flower clusters are smaller and less showy in habit, being more droopy and less horizontal. It is also a vine. *Hydrangea anomala* is also somewhat less cold-hardy than climbing hydrangea, being reliable to zone 5, while *H. anomala* subsp. *petiolaris* is reliable to zone 4.

Neither of these hydrangeas are troubled by diseases or pests. Both do best in moist, well-drained soil that is high in organic matter. *Hydrangea anomala* subsp. *petiolaris* does best with part shade in areas with long, hot summers and could be a good choice for a partially shaded area, such as a porch column. Vegetative propagation is a bit difficult, but cuttings taken in early summer, before the stems brown, have been rooted under mist following application of rooting promoter. Propagation from seed is not difficult; the seed can be sown with no special treatment.

Japanese hydrangea-vine, *Schizophragma hydrangeoides*, has a much smoother, somewhat denser habit than the two *Hydrangea* species. Its leaves are more heart-shaped than the *Hydrangea* vines, and its white flowers are less showy and held in somewhat droopier clusters. Nonetheless, the flowers and glossy foliage are an attractive choice to fill a space in the landscape. This vine is reliably cold-hardy to zone 5, and is more tolerant of hot, wet summers. The cultivar 'Roseum' offers delicate, blush-pink flowers, but the color is very pale and bleaches out in summer heat. Another cultivar of Japanese hydrangea-vine is relatively new and particularly special. Collected in Japan by plantsman and plant collector Barry Yinger, 'Moonlight' (Plate 100) glows with blue-hued foliage as the setting for its soft white flower clusters. This fabulous vine has been distributed by Brookside Gardens in Wheaton, Maryland, and is slowly becoming available commercially.

Schizophragma integrifolium, the Chinese hydrangea-vine, is a close cousin of Japanese hydrangea-vine. Its much larger flower clusters can be 10–12 inches (25–30.4 cm) wide. The foliage of Chinese hydrangea-vine is more oval and less serrated than that of its Japanese cousin but is equally attractive. Chinese hydrangea-vine is especially tolerant of climates with long, hot summers and is less cold-hardy than the other climbing hydrangeas. Chinese hydrangea-vine is not hardy in the north, but will grow throughout zones 7–9. This is one of the best vines for zones 7–9 and probably the best of the climbing 'hydrangeas' in terms of performance in climates with long, hot summers. Unfortunately, *S. integrifolium* is not readily available because of its lack of hardiness, which has inhibited production by the northern commercial nurseries that have traditionally led in the introduction of new plants. Southern nurseries are now beginning to work with this vine, which will be well worth seeking out as it becomes available in the trade.

Culture of *Schizophragma* is similar to that of the climbing *Hydrangea* species. They are also trouble-free in terms of diseases and pests. Propagation of both the Japanese and Chinese hydrangea-vines is accomplished by cuttings. Summer softwood cuttings are treated with rooting promoters and rooted under mist.

Vines offer sculptural opportunities for changing shapes in the landscape. The climbing hydrangeas are ideal for softening harsh structural outlines, or for creating dimension and form around narrow elements such as metal fence posts and chain-link panels. Climbing hydrangea vines add a three-dimensional character that cloaks the garden with the grace of summer.

Sophora japonica / Japanese pagoda tree   PLATE 102

Sophora japonica has an Oriental air about it that brings a contemplative feeling to a landscape. If you listen closely beneath a Japanese pagoda tree in bloom, you'll hear the soft humming of many industrious bees. They are attracted to the tree by the same flowers that catch our own eyes—long panicles of creamy, green-white blossoms that dangle from fine-textured, bright green foliage.

Japanese pagoda tree is native to China and Korea where it was often planted around Buddhist temples, the reason it is still sometimes called "scholar tree." The plant was introduced to America in the 1700s, but it has not been used extensively in the landscape although it is a lovely, undemanding tree.

Sophora japonica grows to 50–70 feet (15.2–21.7 m), casting light, dappled shade from a spreading crown. The creamy flowers appear in July and August in the eastern United States and sprinkle the ground with white as they drop. Flowers mature into interesting cream-colored, chambered pods, which are quite ornamental and dangle from the tree into fall. The refined foliage is divided, providing light shade, a quality that makes the tree valuable for use with lawns, both in parks and suburban settings. The crown of Japanese pagoda tree, unlike many other shade trees, allows enough light penetration to support the growth of turf grass. The fine-textured, light green leaves remain in good condition through the summer. They can develop clear yellow fall color in some years.

While not an extremely rapid grower, *Sophora japonica* has a medium to rapid growth rate and will reach 10–12 feet (3.1–3.7 m)in five to six years. Hardy to zone 4, it is tolerant of many soil types and has few pest problems. It also withstands polluted conditions, making it a good choice for a street tree. Japanese pagoda tree is easily propagated from fresh seed removed from the ornamental pods.

A number of cultivars are available in the trade including 'Pendula', a weeping form that seldom flowers; 'Regent', an early, heavy flowering selection; and 'Violaceae', whose flowers are lightly streaked with a rosy violet color. Cultivars are propagated by grafting.

The quietly beautiful Japanese pagoda tree deserves far more use along our walkways and in our parks and gardens. It is especially useful as a shade tree over lawns because its light, dappled shade allows turf to thrive and the small leaflets do not create a fall maintenance issue. *Sophora*'s semi-pendulous habit makes it a beautiful choice for a shade tree near a pond or other small water feature. The soft white flowers, languidly draped from refined foliage, invite us to rest and reflect in our own garden.

Tamarix ramosissima / Tamarisk   PLATE 103

From a distance, *Tamarix ramosissima* can be imagined as a tall and graceful blue-feathered bird, with rosy plumes waving atop its head, preening above its more plebian garden neighbors. This botanical delight is of course not a bird at all, but a deciduous shrub with very fine, feathery foliage and clusters of tiny, pink-hued flowers that are equally feathery in texture. The tiny, scalelike leaves are pressed close to thin stems and are almost coniferlike in appearance. The delicate foliage is covered with a blue-gray bloom, which gives the plant its blue tone in the garden. The pink- to rose-colored flowers, which are also minuscule, are borne in long, plumed clusters at the tops of the branches in early summer.

Tamarisk is an open, multistemmed shrub reaching 10–20 feet (3.1–6.2 m) in height. Its branches spread out from the crown, much like the tail of a peacock, but in a decidedly more unkempt manner. While the shrub is in leaf and flower, the effect is marvelous, but the plant is best hidden by camouflaging neighbors after it loses its foliage in the fall. Because tamarisk flowers on new growth, it can be pruned back in the late fall or winter to minimize scraggliness and invigorate growth and flowering.

Tamarix ramosissima is native to the saline soil areas of southern Europe and parts of Asia. It is a salt-tolerant plant that is an excellent choice for coastal gardens or any site where salt is a potential problem. Unfortunately, this quality has helped tamarisk to become an invasive alien in the southwestern United States and California, where it should *not* be used as a garden plant. In eastern gardens, however, tamarisk is not as weedy and is a useful garden plant. Tamarisk is hardy as far north as zone 2. It is best in full sun and nutrient-poor, well-drained soils, such as sand, but it also performs well in clay. Too high a nutrient level will result in a leggy, overgrown plant. Tamarisk is subject to powdery mildew toward the end of the growing season. This shrub is readily propagated from fresh seed and can also be rooted from cuttings taken in summer and rooted under mist. In general, the root system of tamarisk is fine and sensitive to being moved. Use container-grown plants to establish new plantings of tamarisk.

The cultivar 'Cheyenne Red' has very deep pink flowers. 'Pink Cascade' is a vigorous plant, with bright pink flowers. The flowers of 'Rosea', a very hardy plant, are rosy pink and appear later than those of other tamarisk selections, blooming as late as mid- to late summer.

There is also another horticulturally available species of tamarisk, *Tamarix parviflora*, small-flowered tamarisk, which is generally a smaller, more demure plant, reaching 10–15 feet (3.1–4.5 m) in height. Its light pink flowers emerge in late spring on old wood, so any pruning must be done immediately after flowering. Other species of *Tamarix* may be encountered from time to time in specialty nurseries or public gardens. They are difficult to distinguish and generally have such similar character and requirements that any tamarisk may be confidently chosen.

Tamarisk catches the eye in the summer landscape, spreading its light blue, feathery foliage and soft pink plumes for all gardeners to admire. It makes an excellent companion to low-growing conifers, both texturally and for winter cover-up purposes. The delicate tamarisk can waft beautifully over and around the evergreen conifers during the spring and summer, and the bare shrub will remain at least partially hidden behind their evergreen foliage during its untidy winter phase. Tamarisk is also beautiful when combined with ornamental grasses or plants with burgundy-colored foliage, such as certain smoketree (*Cotinus coggygria*) cultivars.

Taxodium spp. / Bald cypress   PLATES 104, 105

Deceptively soft in appearance, the unusual deciduous conifers of the *Taxodium* genus are tough and adaptable. Bald cypress, *Taxodium distichum*, is the best known of the group, which also includes the similar pond cypress, *T. ascendens*, and a Mexican species, *T. mucronatum*. The needlelike foliage of *Taxodium* species emerges a bright chartreuse in the spring and expands to create a feathery halo of moss-green around the tree's branches by early summer. Each fall, the foliage turns a lovely bronzed red-gold and drops to create a rusty carpet around the buttressed trunk. The warm cinnamon bark is shaggy, a delightful addition to the winter garden when it can be clearly seen after the leaves have fallen. Rounded cones, 2 inches (5 cm) in diameter, sprinkle the foliage as green ornaments in the spring, turning ruddy brown as they mature.

Taxodium distichum, common bald cypress, is a large tree, reaching epic proportions of 100 feet (31 m) in cypress swamps and other wet areas where trees have remained undisturbed for long periods. The famous bald cypress "knees" develop around the base of the trunk of older plants growing in wet areas. These knobby, bark-covered growths, from 1–3 feet (0.3–0.9 m) high, appear to be aboveground extensions of the roots that may increase aeration in the root tissue—an advantage to the plant in swampy areas. This tough, deciduous conifer is very well adapted to the hot, wet summers of its native Southeast, but it is less well known that *Taxodium* is also very tough and reliable in many other regions and under diverse conditions.

Because bald cypress is generally found in wet areas in the wild, it has often been mistakenly assumed that this tree requires a wet landscape site as well. This is not the case. *Taxodium* seed requires a relatively long period of constant moisture to germinate, and therefore, in natural settings, a wet area is required for *Taxodium* to germinate. But once the seed has sprouted, bald cypress and other *Taxodium* species grow well in average garden conditions and in fact are quite drought tolerant.

Taxodium distichum is found growing naturally from Florida to as far north as Ohio. It is completely hardy to zone 4, and a few older specimens are even hardy enough to continue surviving winter temperatures of –20°F (–26°C). Most landscape plants are seen in the range of 20–30 feet (6.2–9 m) in height with a pyramidal shape and a spread of 10–20 feet (3.1–6.2 m) at the widest point, near the base of the tree, but trees can reach 50–100 feet (15.2–31 m) tall with great age.

Bald cypress requires full sun and relatively acid soil to perform well. Because the tree is deciduous, full winter sun does not create the potential problem for scorch that it can with some other conifers. *Taxodium* can thrive in extremes of both drought and wet conditions, which means it can survive the extremes of clay soils in the summer. There are no serious disease or insect problems.

Taxodium is propagated from seed, which requires three months of moist, cold storage before it will germinate. As with all seed-propagated plants, the provenance of the parent trees will affect cold-hardiness of the seedlings. *Taxodium* is difficult to root from cuttings; hardwood cuttings can be rooted under mist with limited success.

The very few cultivars are propagated by grafting and are quite difficult to find in the trade. 'Pendens' is a pyramidal form with pendulous branchlets. 'Shawnee Brave' is a narrow pyramidal form.

Two other *Taxodium* species, *T. ascendens* and *T. mucronatum*, may be encountered in arboreta and other collections. *Taxodium ascendens*, pond bald cypress or pond cypress, is the subject of much debate among taxonomists as to whether it is a separate species or a variety of *Taxodium distichum*. In the wild, it is generally found in slightly more upland settings than bald cypress, with an overlapping range that is much smaller, including the southeastern states, the coastal plains, and all of Florida. The foliage of *T. ascendens* is remarkably different from that of *T. distichum*. Pond cypress foliage is threadlike and held upright from the branchlets, and it has a blue-gray bloom. *Taxodium ascendens* is more narrowly pyramidal than its cousin. There are two cultivars of pond cypress which are even more difficult to locate than the cultivars of bald cypress. 'Nutans' has short, very horizontal branches; 'Prairie Sentinel' is almost columnar in habit.

Taxodium mucronatum, indigenous to Mexico, is particularly intriguing in the landscape because it holds its sandy gold fall foliage much later than the others, well into December at the North Carolina State University Arboretum in Raleigh, North Carolina. *Taxodium mucronatum* is more broadly spreading in habit and has more open, lighter-colored foliage than other *Taxodium* species.

Mature specimens of *Taxodium* create a refined court of conifers in the landscape. Unusually soft-textured, the trees combine tough adaptability with good character throughout the growing season. These lovely plants are excellent choices for stressful urban uses. They thrive in such inhospitable situations as parking lot planters, streetside strips in the heart of the city, and new developments.

Tilia spp. / Linden

PLATE 106

The lindens are magnificent shade trees with the extra flair of a lovely flowering display. From a distance, *Tilia* in bloom appears to be draped with an antique lace coverlet.

A number of different lindens are in cultivation, ranging from our native *Tilia americana* to European and Asian species, and including an array of cultivars and hybrids. They are all relatively large, eventually broad-spreading, deciduous trees with gray, furrowed trunks and deep grass-green, glossy, oval to heart-shaped foliage that generally does not develop any significant fall color. Though not small, the foliage is handsome, especially on some of the European forms. The shape of the tree is substantial, dense and rounded to conical.

In early to midsummer, depending on the species, lindens are covered with lacy clusters of tiny cream-colored flowers that smell delicious. Flowers eventually mature into interesting nutlets attached to a single winglike, papery bract. The wings start out a delicate light chartreuse and dry to pale beige.

All of the lindens discussed here are completely hardy to at least zone 5, with some having significantly greater cold hardiness as noted. They prefer moist, fertile, well-drained soils but also perform well in clay, and they tolerate a wide pH range. Lindens require full sun but otherwise are adaptable to a wide variety of sites. Most forms are very tolerant of drought, pollution, and urban environments, making them excellent street trees. Propagation is a challenge. Seed germinates with difficulty due to an especially tough seedcoat, and vegetative propagation of cultivars is restricted to grafting. There are a few insect and disease problems on some *Tilia*, including Japanese beetles, which find some lindens irresistible, and powdery mildew, leaf spot, and linden mite.

Tilia is a complex genus, and there is great dispute among taxonomists as to how all of the various hybrids should be classified and whether or not the North American forms should be split into many species. Most horticulturists call all North American *Tilia* "*T. americana*" (with the exception of *T. heterophylla*, white basswood, which is still separated by some, especially in its native South).

Tilia americana, basswood or American linden, is indigenous to the entire eastern and middle sections of North America, from the Great Lakes and St. Lawrence Seaway areas of Canada to Alabama and points between. It is hardy to zone 2. It is the largest and most coarse of the lindens in use horticulturally, reaching 80–100 feet (24.8–31 m) with age, with leaves 4–7 inches (10.1–17.7 cm) long and nearly as wide. It has a more open growth habit than other forms. Named selections are handsome and tough and are excellent trees for streets, parks, or large properties. 'Dakota' is a very rounded selection; 'Douglas' is a broadly pyramidal, upright form; and 'Fastigiata' has a formal, narrowly pyramidal habit. 'Legend' is one of the best cultivars with excellent quality foliage late into the season and good, semiformal habit. 'Rosehill' is one of the fastest growing basswood selections.

Tilia heterophylla, white basswood, has been lumped with *T. americana* by taxonomists. It differs from American linden in its leaves, which have distinctly silver undersides that are quite beautiful. White basswood is native to the southern part of the United States and is hardy to zone 5. One named selection of *Tilia heterophylla* is 'Continental Appeal', which has very silvered foliage and formal, narrow, upright habit. It is an excellent choice for an urban street tree.

Tilia cordata, littleleaf linden, is the historical linden of Europe that has been important in horticulture there for centuries. Its leaves are very neat and small, giving the tree greater elegance in the landscape than its cousins. In Europe, it has been used to create sheared hedges that are dramatically effective; this technique could be used more often here as well. Littleleaf linden is hardy to zone 3. There are a myriad of cultivars. 'Bicentennial', selected in the United States, has a tight, upright habit. 'Erecta' (or 'Bohlje') has a fastigiate, narrowly columnar habit that makes it a good street tree in tight urban spaces. It has the added advantage of developing respectable yellow-gold fall color. 'DeGroot' has lovely, glossy foliage and slower growth than others and is a good choice for smaller scale landscapes. 'Glenleven' is a rapidly growing selection out of Canada. Its habit is more open than other littleleaf forms, more like basswood. The aptly named 'Green Globe' forms a tightly rounded crown; it is generally top-grafted onto taller understock to ultimately form a specimen 15–20 feet (4.5–6.2 m) tall with a globe-shaped crown. 'Greenspire' is probably the most widely grown cultivar. It has excellent foliage and good branching, and gives reliable long-term performance in tough sites. 'Handsworth' is an interesting form; its young stems are a light chartreuse color. 'Morden' is a slow-growing cultivar with neat, formal outline. 'Pendula Nana', a dwarf form with somewhat pendulous branch tips, would make an interesting container tree for enclosed spaces. 'Rancho' is especially floriferous; the blooms have an intense yet pleasant fragrance.

Tilia tomentosa, silver linden, is native to Europe and western Asia. It is named for its two-tone foliage, which is deep glossy green on the upper surface and silvery white on the underside. The lower surfaces get their color from a proliferation of small hairs (a characteristic similar to some of the poplars). The eye-catching effect is almost magical in a breeze, as the leaves turn this way and that, revealing the bright silver underneath. Silver linden has very smooth, light gray bark, reminiscent of the bark of beeches (*Fagus* spp.). The tree is a bit smaller and rounder than *T. cordata* and *T. americana*. It is hardy to zone 4. Some good cultivars are 'Erecta' (or 'Fastigiata'), which is upright and more narrow than the species; 'Green Mountain', a heat-tolerant selection from Princeton Nurseries, New Jersey, that is a good choice for areas with hot, wet summers; and 'Sterling', which has remarkably silvered leaves.

Other species from Europe include *Tilia × euchlora*, Crimean linden; *T. × europaea*, European linden; and *T. platyphyllos*, bigleaf linden. *Tilia × euchlora*, Crimean linden, hardy to zone 3, is smaller than *T. americana*, our native linden, reaching 40–50 feet (12.4–15.2 m) in height. It has a less formal, more graceful habit than *T. cordata*, littleleaf linden. *Tilia × europaea*, European linden, on the other hand, is a large tree that can reach 100 feet (31 m) in height. It has been used all over Europe to create allées and street plantings. It suckers readily from the base. It is hardy to zone 3, but can have severe insect problems in the United States from aphid and spider mite infestations. *Tilia platyphyllos*, bigleaf linden is one of the parents of *T. × europaea*. It has exceptionally large leaves for a European linden. There are some unusual cultivars of *T. platyphyllos*, including 'Aurea', with yellow young twigs; 'Laciniata', with lobed leaves; and 'Tortuosa', with contorted branching. It is hardy to zone 4.

Two species with Asian roots are *Tilia mongolica*, Mongolian linden, and *T. petiolaris*, pendent silver linden. The beautiful Mongolian linden is a small tree reaching 30–40 feet (9–12.4 m) in height with the appealing character of distinctly cut leaves. It is hardy to zone 3. *Tilia petiolaris*, pendent silver linden, is very similar to *T. tomentosa*, silver linden, but with nearly weeping branching. Hardy to zone 5, *T. petiolaris* can develop very good, clear yellow fall color. There are a few reports that the flowers of this species are "narcotic" to bees, causing them to fall to the ground under the tree.

It is certainly clear that linden is royalty among shade trees. The many forms of *Tilia* offer handsome form and exquisite, glossy foliage. They are the darlings of landscape architects because of their formal, predictable outlines, but they make lovely single specimens on rolling properties as well as excellent suburban shade trees. Their floral display is not often remarked upon, but it is quite noteworthy. The sweetly scented, lacy blooms add another element to this beautiful shade tree.

Ulmus alata / Winged elm   PLATE 107

The tragic introduction and spread of Dutch Elm Disease (DED) wreaked havoc with the famed arching green canopies of _Ulmus americana_, American elm, which once dominated many city streets. Only a few remnants persist of the American elms that once covered the eastern half of the United States. In the wake of DED, an army of breeding and research programs has attempted to develop a new selection of elm that will exactly replace _U. americana_. Sadly, this has proven to be essentially impossible, in spite of the selection and release of many outstanding new hybrid cultivars with excellent DED resistance and landscape character. None seem able to bring quite the same stately grace to our streets, parks, and landscapes.

The increased attention on alternative elm species from around the world has revealed excellent trees, not only for use in breeding programs, but also as potential landscape plants in their own right. One of the best of the "other elms" is the underappreciated but superb _Ulmus alata_, winged elm, an indigenous American elm.

Ulmus alata is native to or naturalized in the southeastern and south-central United States. It can be found growing in the Piedmont and foothills, usually in moderately dry areas, such as old fields and upland woods, but occasionally it is seen in low, wet areas as well. Winged elm is named for its habit of developing flattened, corky outgrowths from the bark of young branches and twigs, similar to that of sweetgum, _Liquidambar styraciflua_, or winged euonymus, _Euonymus alata_. A small to medium-sized deciduous tree, _U. alata_ reaches 20–40 feet (6.2–12.4 m) in height with a spread of 15–30 feet (4.5–9 m). The branches are fine-textured, but they are held at rather broad angles to the trunk and so create an open, airy canopy. In combination with the winged bark, the effect is beautifully architectural without becoming overly coarse or visually demanding, and the result can be stunning when the leaves have fallen in the winter garden. In the wild, the degree of branch winging is extremely variable among seedlings, with some showing magnificent wings 2 inches (5 cm) wide, and others with merely sporadic bumpiness on the youngest branches only. This variability is an excellent reason to procure vegetatively propagated trees from a well-winged parent tree.

The leaves of winged elm are small, finely serrated, emerald-green ovals with noticeable veining that lends a delicate, lacy texture to the tree's appearance. Leaves develop a subtle, pale yellow fall color which is very restful, and the tree is a good foil for plants with more dramatic fall color. Foliage retains excellent quality throughout the hot summers with no significant pest or disease problems. The tiny, inconspicuous flowers of winged elm mature into clusters of interesting russet-colored, diminutive winged seeds. The seeds are ringed with barely visible whitish fringe so that they glow when backlit by the setting sun.

Ulmus alata is completely hardy to zone 6. It prefers moist, well-drained soil in full sun to partial shade but will tolerate a range of soils, from heavy clay to drier, sandier soil. Winged elm is propagated from early summer softwood cuttings, rooted under mist. For best success, cuttings are taken from the most juvenile area of the tree accessible (the lowest branches). Winged elm is also propagated from its mature seed, which requires three months of storage in a moist medium, such as peat, at cold temperatures of about 40°F (4°C).

No named selections are currently available in the trade; however, the North Carolina State University Arboretum has been extremely fortunate to receive a donation of a unique seedling variant—a 45 year old, weeping winged elm, 8 feet (2.4 m) tall and 12 feet (3.7 m) wide. This elegant tree combines the architectural grace of a Japanese maple with the fascinating texture of its native cousins and has been a dramatic addition to the Arboretum's collections. It has been propagated and distributed to commercial nurserymen and should be available in nurseries in a few years, under the name 'Lace Parasol' (Plate 107).

Winged elm is an elm of superior beauty and character and is deserving of far greater use and attention than it currently receives. It makes an excellent small tree for a mixed shrub border, and it is the appropriate size for a shade tree in a suburban garden. On large properties, it can be planted en masse, as a surprising grove. Winged elm is a very tough elm that may also have good potential as a street tree in the small spaces of urban thoroughfares. The winter branch profiles are spectacular against stone walls, buildings, and fences, especially with night lighting. The winged branches also make fine additions to cut arrangements, offering intriguing texture that does not overwhelm.

Vitex spp. / Chastetree   PLATES 108, 109

Two large shrubs or small trees, *Vitex negundo* and *V. agnus-castus*, share the common name of chastetree, and both offer gentle gray-green, lacy foliage, decorated with lavender spires of tiny, delicate flowers. Both species are deciduous, multistemmed plants that reach 10–15 feet (3.1–4.5 m) high, with arching branching that spreads into a rounded canopy at the top. Both have hand-shaped, divided leaves with very narrow "fingers." The divided character of the foliage gives the plant its lacy appearance. The leaves are usually a refreshing gray-green color in spring and summer, with little or no development of fall color.

These natives of southern Africa and the Philippines are great lovers of heat and perform well in any full sun site with regular moisture. Both chastetrees prefer moist, well-drained soil, but both also do well in relatively heavy clay. There are no serious pest or disease problems with either chastetree species. Both are easily propagated from softwood cuttings, taken before flowering and rooted under mist, or by seed, which needs no pretreatment.

Vitex negundo is the larger and more hardy of the two. While it is completely hardy to zone 6, it cannot be grown further north with any certainty of avoiding winter kill of the aboveground parts of the plant. The flower clusters are spikes of tiny lavender flowers, reminiscent of the flowers of butterfly bush (*Buddleia* spp.), but much smaller and more delicate. 'Heterophylla' is a beautiful cultivar with fernlike lobing on the leaves, giving it an extra-lacy look and especially refined texture. 'Heterophylla' is significantly more hardy than other *Vitex negundo*, with no damage down to −10°F (−21°C), making it the best choice for gardens in the colder parts of zone 6.

Vitex agnus-castus is often a somewhat smaller plant than *V. negundo*, but its flowers are more showy, with larger flower clusters, 6–12 inches (15.2–30.4 cm) in length, in shades of blue-violet, pink, or white. It is somewhat less cold-hardy than *V. negundo*, being reliable to zone 7. 'Rosea' is a lovely pink-flowered form. 'Alba' and 'Silver Spire' produce bright white flowers.

Both *Vitex* species are versatile plants. They are excellent both limbed up as small trees or left to develop as multistemmed large shrubs for use in mixed shrub plantings. By continually pruning the plant to 1 to 3 main stems, and selectively removing side branches, you can train the plant to develop into a small tree. Groupings of *Vitex* make beautiful, airy massed plantings.

Chastetree can be treated as a woody perennial, in a similar fashion to *Buddleia*, by cutting the entire plant back to within 1–2 feet (0.6–0.9 m) of the ground. This technique will stimulate lush vegetative growth the following season, and since *Vitex* flowers on new growth, there should be a good floral display as well. *Vitex* combines floral and textural interest in the garden when few other woody plants are in bloom. The cool colors of the flowers above the silvery foliage are especially appealing as the heat of summer builds.

Fall

Abelia × *grandiflora* / Glossy abelia   PLATE 110

"Do you have a shrub that will grow in any soil, take sun or shade, flower for months through spring, summer, and fall, grow quickly but not get too large, make a hedge if I want one (or stay neatly in its corner), babysit the children, not mind the neighborhood dogs, and, oh yes, require little care or maintenance?"

How many times have garden-center staff heard this question (or some version of it) from new homeowners, busy landlords, or anyone with a difficult landscape site? The amazing thing is that, believe it or not, there actually is a shrub that will do all of these things.

Abelia × *grandiflora*, the glossy abelia, displays the best qualities of its two parents, *A. chinensis* and *A. uniflora*. It is a dense, but somewhat fine-textured, multistemmed shrub reaching 4–6 feet (1.2–1.8 m) in height with a nearly equal spread. The branches arch up and over, creating a soft, vase-shaped outline. The bright green foliage of glossy abelia is indeed exceptionally glossy, even as it turns a beautiful wine red through the winter months. The small, pointed, oval leaves, only about 1 inch (2.5 cm) long, densely cover the branches. Glossy abelia is evergreen throughout most of zone 7 and warmer, but it will drop foliage in colder areas of the southern mountains.

Glossy abelia flowers continuously from spring throughout the summer and into fall with large numbers of perky, pale pinkish, 1-inch-long (2.5 cm) blossoms scattered across the ends of all of the branches, sprouting from the shiny green foliage. It is delicately pleasing in flower and lacks the overwhelming quality of its close relative, forsythia (*Forsythia* × *intermedia*), because the flowers are not borne in a solid mass.

Abelia × *grandiflora* will successfully grow in almost any site that is not extremely wet. It prefers moist, well-drained soils in full sun but will be quite handsome in moderate shade and tolerates a range of soils from dry to poorly drained heavy clay.

Glossy abelia is not a good choice for sites with extreme winters. It is completely hardy to zone 6 and zone 5 where temperatures remain above approximately −5°F (−18.5°C). Temperatures below −5°F (−18.5°C) may result in die-back and stem damage, although the roots will usually survive if exposure to these temperatures is infrequent and short-lived. Glossy abelia can be pruned back severely in late winter to stimulate new growth and flowering for the following spring. It has no significant disease or insect problems.

Abelia × *grandiflora* cultivars are propagated from softwood cuttings taken in the spring and summer and rooted under mist. The selection 'Confetti' boasts spectacular pink and white variegation. 'Francis Mason' has gold-yellow new foliage. 'Prostrata' is a compact, lower growing form. 'Sherwood' is also a compact form, with even smaller leaves than the original hybrid. 'Variegata' is a rare variegated form. 'Edward Goucher' is a hybrid of glossy abelia with yet a third abelia species, *A. schumannii*, and is considered by many to be the best garden form of abelia. The deep lavender-pink flowers of 'Edward Goucher' are somewhat larger and showier than those of glossy abelia and are very striking against the dark green, shiny foliage. 'Edward Goucher' is significantly less hardy than glossy abelia and will not do well where temperatures dip to 0° to −5°F (−16° to −18.5°C). At the North Carolina State University Arboretum, 'Edward Goucher' makes an especially lovely addition to the mixed shrub border.

The beautiful flowers, foliage, and form of glossy abelia make it an excellent choice for hedges, low screens, massed plantings, or even as a single solution for a difficult corner. The shape and texture of the shrub combine with its charming flowers to make *Abelia* × *grandiflora* an easy plant to use in a wide range of landscape and garden settings. The marvelously long blooming season offers dependable transitional interest from spring to summer and into fall as other more showy plants pass in and out of peak bloom.

Abies firma / Japanese fir   PLATES 111, 112

The firs of the mountains are striking plants in cooler areas, but they generally die out in warmer landscapes due to root rot or related disorders. This does not mean that gardeners in the hotter regions of the country must suffer through gardening without any firs at all. There is a fir which will thrive and grow into a magnificent specimen, even in the hot, wet climates and clay soils that will kill other firs.

Abies firma, the Japanese or momi fir, is native to Japan, where it is the tallest and most arresting fir in its native habitat, eventually reaching 150 feet (45.7 m). In cultivation, it will probably reach anywhere from 20–50 feet (6.2–15.2 m) in height with age, but it is rather slow-growing. The light emerald-green foliage is more stiffly erect and sharply edged than that of most firs. The ends of the short, flattened needles are distinctly pointed. The tree's habit is conical with somewhat open branching. The bark is a handsome light gray. The yellow-green cones mature to a light tan and may reach 5 inches (12.7 m) in length. The cones are borne upright on the branches, as they are on all firs and true cedars (*Cedrus* spp.).

Japanese fir is not quite as cold-hardy as other firs, being reliably hardy to zone 6. It may suffer some dieback in areas with severe winters, but it thrives in areas with hot, humid summers where other firs may languish.

Abies firma is an excellent way to introduce the beauty of a fir to gardens inhospitable to other firs. The name *firma* means "strong," and this is indeed a tough plant. It performs well in heavy clay soils and is tolerant of a range of moisture conditions. It requires full sun to prosper. It is best to give it the most optimal drainage available on your site, but don't hesitate to plant this tree if heavy clay is the only soil in your garden. Japanese fir is especially tolerant of hot, wet summers. This tree will continue to thrive through weeks of 80°F (24°C) nights, conditions under which other firs, such as the popular balsam (*A. balsamea*) or northern white fir (*A. concolor*), would soon die. At the North Carolina State University Arboretum, a Japanese fir thrives in the heavy soil of the west arboretum, just west of the Japanese garden, illustrating the versatility of this tough plant.

Like all pine relatives, Japanese fir is difficult to propagate vegetatively and must be grafted onto seedling rootstock. The species is propagated from seed and is very slow to develop to a size large enough for planting out in the landscape. Cultivars are essentially unavailable in North America, but the species can be found with a bit of searching.

Like all firs, *Abies firma* makes an excellent evergreen specimen with a relaxing formal silhouette. However, it has a softer, more irregular outline than other species, which makes it easier to combine with other plants in the garden. It is particularly handsome when used in conjunction with low walls or benches and near rock or stone. Japanese fir combines the stately elegance of firs with good adaptability to the demanding conditions of warm climates and heavy soils. It is an exceptional choice for bringing dramatic structure and beauty to gardens in zones 6 to 9.

Acer leucoderme / Whitebark maple   PLATE 113

Whitebark maple, a native of the southern United States, is a close relative of the more northerly sugar maple (*Acer saccharum*). It brings vibrant fall color and other attributes similar to those of its northern cousin to the southern landscape, yet it thrives in a warm climate. *Acer leucoderme* also offers interesting pale gray or chalky white bark (*leucoderme* means "pale skin") on the upper part of the tree.

Both sugar and whitebark maples have oval to rounded crowns that open as they mature, but whitebark maple is a more petite version of sugar maple. It reaches mature heights of 25–30 feet (7.5–9 m), as opposed to the 50–70 feet (15.2–21.7 m) of height that mature sugar maples can attain. Whitebark maple's leaves are also smaller in size. In the fall, however, *Acer leucoderme* is nothing less than the equal of its larger Yankee relative. Fall color ranges from brilliant yellow-orange to cardinal crimson.

The whitebark maple is a United States native that grows as an understory tree, like dogwoods and redbuds, in well-drained, upland woods of the southeastern United States, especially Georgia and the Carolinas. In the early fall, whitebark maples create pools and flashes of vivid color in the southern woods. No other tree can rival their showiness at that time of year.

Acer leucoderme is cold-hardy in zones 5 and warmer. It prefers well-drained sites and will tolerate dry areas. It is essentially free of insects and diseases. Whitebark maple can be grown in a range of light conditions, from full sun to heavy shade. Full sun is necessary to give the best fall color.

While not common in most garden centers, *Acer leucoderme* is worth seeking from native plant nurseries. Whitebark maple is propagated from seed collected in fall and sown outside in soil. Seedlings will germinate and grow the following spring. Year-old plants can be easily transplanted to their permanent sites.

Whitebark maple makes a lovely, multistemmed lawn or shade tree. It is especially useful for urban and suburban residential landscapes because of its relatively small stature and good heat tolerance. Planted as a single specimen or in a small grouping, it will give dramatic color to the autumn garden.

Acer triflorum / Three-flower maple   PLATES 114, 115

Fall color is the topic of conversation throughout much of North America from September through November, but the deciduous plants that offer this grand display are not concerned with their own art—they are working to survive the winter extremes. Their annual apparent change in foliar pigments is a sideshow to the main act of survival.

As gardeners, we take advantage of this adaptive behavior by planting trees and shrubs with the most vibrant of fall displays. Gardeners tend to rely on the familiar for fall color, neglecting many excellent plants with brilliant fall color. One example of such an underused, excellent tree with spectacular fall color is *Acer triflorum*, three-flower maple. The young grass-green foliage of three-flower maple is elegant in spring and early summer, but in autumn, its divided foliage becomes a dazzling spectacle of oranges, scarlets, purples, and golds.

Acer triflorum is a small deciduous tree, native to Korea and Manchuria, that will ultimately reach 20–30 feet (6.2–9 m). In cultivation, it is usually seen in the range of 12–15 feet (3.7–4.5 m), similar in size and proportion to many Japanese maples (*A. palmatum*). It is one of the group of "trifoliate" maples, which all have leaves divided into three long lobes.

Fall color is the primary reason that *Acer triflorum* has been grown, but this tree also offers magnificent peeling bark, even as a young plant. The bark is warm light brown in color. The flowers are also of interest. They emerge in pinkish clusters of three, giving rise to the common name of "three-flower maple."

This remarkable plant will grow well in full sun, where it is rounded, compact, and full, or in part shade, where it is upright and spreading (again, like many Japanese maples). Fall color on a small tree is well developed, even in the shade. Three-flower maple prefers moist, well-drained, acidic soils, but it also performs well in heavier soils, and is drought tolerant once established. It is also relatively cold-hardy, being reliably hardy through zone 5 and surviving to zone 4. No serious pest or disease problems exist.

One of the reasons *Acer triflorum* is not commonly available is because of its difficulty of propagation. Seed often has low viability and has a double dormancy requirement, needing both warm and cold periods before it will germinate. The seed also has an exceptionally hard seedcoat that requires scarification in addition to the temperature treatments. Vegetative propagation is not much simpler. Cuttings have low rootability, and so plants are often grafted, frequently onto paperbark maple (*A. griseum*) rootstock, which is not well-adapted to hot climates and heavy soils. If you garden in such areas, beware of grafted plants.

Three-flower maple is one of the finest small trees for modern landscapes. Gardeners are always searching for handsome small trees with multiseason interest that are well adapted to tough conditions. Three-flower maple is just such a tree. Its moderate growth rate keeps it a manageable size for small properties and tight spaces. Fall is the perfect time to include this plant, at the peak of color, in our own autumn gardens.

Aronia arbutifolia / Red chokeberry   PLATES 116, 117

Autumn is the most dramatic season of the year for red chokeberry, *Aronia arbutifolia*, a native deciduous shrub. The leaves turn rich shades of scarlet and wine-red, and the fruits develop into bright crimson berries. This fall color is especially striking when the shrub has been planted en masse.

Red chokeberry grows 6–10 feet (1.8–3.1 m) tall and can spread to 10 feet (3.1 m) wide. It is native to much of the eastern half of the United States, from New England to Florida and west to Ohio and Texas. It is an upright, multistemmed shrub that readily suckers and spreads over time to form a colony. *Aronia arbutifolia* tends to get leggy as it matures because foliage is lost from the lower part of the plant. Planting the shrub in masses or hedges will eliminate any visual unattractiveness due to this character.

The four-season interest of red chokeberry begins in the spring with delicate, white to pinkish flowers not unlike those of the serviceberry (*Amelanchier* spp.). In summer, the dark, glossy green foliage is very attractive and has the added interest of a soft gray, feltlike covering on the leaves' undersides. The fruit ripens in fall and persists into winter. The cheerful red berries on the brown stems add a wonderful dimension of color to the gray tones of a dormant garden. The berries have been used to treat scurvy because of their ascorbic acid content, but they are very astringent, so bitter that even the birds will not eat them—hence, the name "chokeberry."

Aronia arbutifolia is tolerant of a wide range of growing conditions from low, wet soils to poor, dry ones. Full sun will give the best fall color and fruiting, but the plant will do well in partial shade as well. It is hardy through zone 4. Red chokeberry can be propagated by seed, softwood cuttings, or division of existing colonies. 'Brilliantissima' (Plates 116, 117) is an excellent cultivar of red chokeberry with more prolific flowering and fruiting than the species and exceptional fall color as well.

Planting red chokeberry in a mass or grouping will create a cove of landscape beauty year-round. From spring's sweet white blooms and summer's deep green gloss to fall's flash of brilliant red and winter's touch of crimson, *Aronia arbutifolia* brings its own unique character to all the seasons of our gardens.

Baccharis halimifolia / Groundsel bush   PLATE 118

Along the highways and byways of the American east coast, near the end of the beach season, a soft haze of silver appears, tucked among the autumn grasses and undergrowth. This feathery show is put on by the fruits of groundsel bush, *Baccharis halimifolia*. Each fall, after flowering, this shrub is covered with thousands of little fruits resembling tiny white paintbrushes. Each paintbrush is actually a collection of silky hairs, or pappus, that surrounds every individual fruit. Together, the pappi make a striking display that looks like a cloud of angel hair on each shrub.

Baccharis halimifolia is a deciduous shrub native to coastal areas of the southeastern and middle-Atlantic United States and extending inland in the Southeast. It is one of the very few shrubs in the Composite family of mostly annuals and herbaceous perennials. The foliage is a handsome silvery gray-green through the growing season. The leaves do not develop noticeable fall color and the flowers are nothing to speak of, but the silvery white pappi on the fruit add a fairylike dimension to a mixed border or naturalized area.

Groundsel bush grows rapidly to reach 5–12 feet (1.5–3.7 m) in height. It is an open, airy-textured plant that spreads to 10 feet (3.1 m) wide with a loose, irregular shape. *Baccharis halimifolia* is a native "pioneer" species. Once found only near coastal areas from Massachusetts to Florida, it has successfully colonized inland regions as well. Groundsel bush is quite cold-hardy, to zone 5, and will tolerate most growing conditions. It is a good choice for coastal landscapes, because it thrives in full sun and is very salt-tolerant. It also makes a fine garden plant in inland regions, tolerating heat, drought, and poorly drained soils.

In spite of its many virtues, *Baccharis halimifolia* is rarely seen in commercial nurseries; only a very few native plant nurseries produce the plant. The species is dioecious, with male and female flowers are on different plants. Seed collected from female plants in the fall needs no special treatment and can germinate in one to two weeks. Seedlings grow readily, and when female plants and male plants are grown in close proximity, many seedlings can be produced. Groundsel bush can also be propagated vegetatively from softwood cuttings taken in summer from selected superior plants.

Groundsel bush is an unusual addition to a perennial or mixed shrub border, painting the fall landscape with its lovely, silvery white brushes. It is particularly appealing when planted with perennials that are its companions in its natural habitat—the fall-blooming asters (*Aster* spp.), goldenrods (*Solidago* spp.), and ironweeds (*Vernonia* spp.).

Callicarpa spp. / Beautyberry   PLATES 119, 120

Callicarpa means "beautiful fruit," and beautyberries are at their best each fall when their berries mature to intense violet or metallic lavender-blue, like a splash of crayon color decorating the stems of an otherwise unassuming, deciduous shrub. Some beautyberries also develop a rosy tinge in their leaves in the fall, which contributes to the overall display, but it's that electric grape color of the fruit clusters that makes beautyberry a favorite for bold display in the fall garden.

Beautyberries are generally coarse and unremarkable throughout the rest of the year. They flower in spring with light pink or pale lavender flower clusters. The flowers are not especially showy but are delicately attractive. They mature into the fabulous fruit clusters. The fruit generally persists after leaf-fall, which makes these shrubs an excellent choice to add interest to evergreen plantings or anywhere late fall displays are desired. There are also white-fruited forms of almost all of the beautyberries, which can be successfully interplanted with the purple-fruited forms or used on their own as less obtrusive plants for fall interest. The white berries are especially effective against broadleaved evergreens.

There are a number of species of beautyberry, which range in height, hardiness, foliar character, and refinement. *Callicarpa americana*, the American beautyberry or French mulberry, is native to the southeastern United States and southwest into Mexico and also in the West Indies. This is the largest and most coarse of the beautyberries, reaching 5–8 feet (1.5–2.4 m) tall with an equal to greater spread. Very large leaves, up to 6 inches (15.2 cm) long and half as wide, are borne on very loose, open stems. The foliage is fuzzy and a light, rather yellowish green. Not as hardy as other beautyberries, *Callicarpa americana* is reliable only through zone 7. It needs probably the most moisture of the beautyberries, doing best in moist, well-drained soils, but tolerating poorly drained clays as well. It gives the best fruit show in full sun. The fruit is borne tightly pressed against the stem, in clusters around the leaf bases (Plate 120). This shrub can get quite large and ungainly, but since it flowers on new growth, it can be pruned to within 6 inches (15.2 cm) of the ground in early spring or late winter. *Callicarpa americana* var. *lactea* is a white-fruited form of the species.

Probably the most widely grown garden beautyberry is *Callicarpa japonica*, the Japanese beautyberry. Native to Japan, as the name suggests, this is the hardiest beautyberry, being reliable to zone 5. Japanese beautyberry is somewhat more refined than its American cousin but is still relatively coarse. It will reach 6–8 feet (1.8–2.4 m) in height and spread. There is some argument among horticulturists as to whether it is as showy in fruit as some of the other beautyberries, especially since its fruit are borne in loose, open clusters. *Callicarpa japonica* prefers well-drained soils and, like its American cousin, does best in full sun. Pruning to within a foot (0.3 m) of the ground in spring is a good way to rejuvenate plants that are leggy (it flowers on new wood). The foliage of Japanese beautyberry can achieve an almost lavender fall color that is beautiful with the intensely purple fruit. The handsome cultivar called 'Leucocarpa' is a particularly attractive white-fruited beautyberry.

The Bodinier beautyberry, *Callicarpa bodinieri*, a native of China, is very popular with English gardeners and deserves a place in American gardens as well. Its fruit are smaller and more refined than other species and are quite shiny. Like Japanese beauty-berry, the shrub is moderately coarse-textured and will also reach 6–8 feet (1.8–2.4 m) at maturity. It has bluish fruits, borne in somewhat open clusters. The fruits do not persist as well those of American beautyberry. *Callicarpa bodinieri* is cold-hardy through zone 6, but it may not be as tolerant of hot, wet summers as *C. americana*.

Callicarpa dichotoma, the purple beautyberry, is an introduced native of China and Japan considered by many to be the most refined and graceful beautyberry for gardens. It is smaller than other beautyberries, reaching only 3–4 feet (0.9–1.2 m) in height with a 4–5 feet (1.2–1.5 m) spread. The foliage is darker, smaller, more dense, and generally more handsome than other beautyberries. The fruit is bright lilac and usually prolific, giving a lovely display. *Callicarpa dichotoma* var. *albifructus* is the equally attractive white-fruited form.

At the North Carolina State University Arboretum, many of the callicarpas surprise visitors with their fall displays. The *Callicarpa* shrubs in the early–late border at the west end of the perennial border, an American beautyberry and a white-fruited Japanese beautyberry, are beautiful examples of *Callicarpa* in a gardenesque setting.

All of the beautyberries bring a special bold beauty to fall gardens. Their brightly colorful fruit is a beacon for the eye and a delight for any gardener. The display is particularly effective when the shrubs are planted in groups. The bright crayon colors of the uninhibited display of berries add a wonderful whimsy in the autumn garden.

Choisya ternata / Mexican orange   PLATE 121

The early fall bloom of Mexican orange, *Choisya ternata*, is a special surprise at a time of the year when the mood of the garden turns toward the bronze and red colors of falling leaves. A member of Rosaceae, the same botanical family as roses, Mexican orange bears flowers with a similar, though less intense, fragrance than its more familiar sibling. Although their fragrance is rosy, the flowers look very much like the flowers of true orange.

Mexican orange is an evergreen shrub with a pleasingly round and mounded habit. It is not hardy beyond zone 7. It will eventually reach heights of 8 to 10 feet (2.4–3.1 m) with an almost equal spread. The foliage is beautiful all year where it is hardy, retaining its shining elegance through the winter. The deep green, glossy leaves of this shrub are so glossy and perfect that some garden visitors wonder if they are plastic! The leaves are a perfectly archetypal leaf shape—about 3–4 inches (7.6–10.1 cm) of smooth oval tapering to a pointed tip—and are arranged in groups of three like small, three-fingered hands. The bright white flowers are borne in the greatest profusion in late spring, but there is often a secondary bloom in the early fall and sporadic flowering can occur most any time of the year when the temperatures are not extremely hot or cold.

Choisya ternata prefers well-drained, moist, slightly acidic soils in full sun or partial shade. It also performs well in lighter clay soils. It is propagated by softwood cuttings taken in summer and rooted under mist.

'Sundance' (Plate 121) is an exciting cultivar with sunflower-gold foliage. The new foliage is the brightest gold. It is most colorful during the late winter, making it a superb choice to brighten winter gardens.

This shining shrub, an excellent alternative to the usual broadleaved evergreens, makes a lovely specimen plant or a massed grouping in the landscape. The delicate, fragrant white flowers, sprinkled among shining foliage, are always a surprise in the fall. Their understated display reminds us of the effectiveness of subtlety in the garden.

Diospyros virginiana / Persimmon   PLATE 122

Persimmon trees produce persimmons, of course, which some consider a mixed blessing. In the early autumn, the orange to pink fruits are attractive on the tree. When fully ripe, they are delicious eaten fresh or in puddings—but beware the unripe persimmon! The small, soft orange fruit peeking through dark green foliage look especially inviting. Giving in to temptation, you reach for one; it resists leaving the tree but comes off in your hand. A big bite and then—Oh dear!—more pucker power than any sour lemon ever thought of.

Unripe persimmon fruit are notoriously astringent, and truly ripe fruit still on the tree are hard to distinguish from their unfriendly neighbors. In spite of the tricky fruit, our native persimmon tree, *Diospyros virginiana*, has much to offer American landscapes. The trees have a picturesque shape that's well suited to naturalistic settings, and the unique checkered bark is reminiscent of alligator skin. Persimmon trees generally have attractive yellow to purple-bronze fall color that provides a good backdrop to the decorative fruit, and even after leaf-fall, the persistent fruit are ornamental.

Diospyros virginiana is native to moist bottomlands as well as dry sites and uplands in the Southeast, southern Midwest, and middle-Atlantic areas of the United States, from New Jersey to Florida and west to Indiana and Texas. It is often seen in abandoned fields, along roads, and on low-fertility soils.

Persimmon is a member of the ebony family, an interesting group of mostly tropical, hard-wooded trees that also includes the ebony tree. Persimmon wood is very dense and close-grained and is used for golf-club heads, veneers, and billiard cues.

Persimmons are deciduous trees with an irregular habit and a somewhat slender trunk for their height. Relatively slow-growing, the tree can reach 35–50 feet (10.9–15.2 m) in height with an oval crown 20–30 feet (6.2–9 m) wide. The foliage is dark green and lustrous with a lighter underside. One of the most interesting and distinctive aspects of this tree is the dark gray bark, which covers persimmon trees with a thick hide of small, squarish blocks. Children love the feel of it.

The bell-shaped, creamy colored flowers are inconspicuous but mature into attractive fruit. The fruits are small and round, about 1–2 inches (2.5–5 cm) across, with 4 persistent and noticeable remaining floral parts where the fruit is attached to the stem. These flattened lobes dry and extend out from the fruit like little wings, but are only about 0.5 inch (1.2-cm) long. The fruit takes a long time to ripen and requires several frosts to soften and sweeten and lose its famous "pucker power" (although some cultivars have been selected for good fruit ripening without the need for frost). Fruits are scattered on the twigs and stems throughout the crown of the tree.

Persimmon trees prefer moist, well-drained soils but will do fine in dry, poor soils as well. They are hardy to zone 4. Established persimmon trees are difficult to move because they have long taproots. Purchase balled-and-burlapped young trees and transplant in the spring to avoid difficulties. Black leaf-spot may occur on persimmon trees late in the season, but the trees are not usually severely affected and any damage is mostly cosmetic.

Persimmons easily reproduce themselves from seed. The trees are often found growing along fencerows and country roads, where the seeds have been dropped by wildlife. Persimmon is readily propagated by stratified seed. The seed is layered in peat moss or some other moisture-holding material, such as sand, and, usually, stored for some time at cold temperatures. This treatment helps break the seeds' dormancy. Cultivars are propagated asexually and grafted onto seedling understock to retain their unique character. Seedling trees vary greatly in fruit size, quality, color, and taste. To assure dependably high quality from planted landscape trees, use named varieties of persimmons. Because seedling trees can be either male or female, it is especially useful to purchase named cultivars (whose sex is thus known) to insure fruit production. Some good female cultivars are 'Early Golden', with prolific, early fruit; 'Morris Burton', with reportedly some of the best-tasting fruit; and 'Wabash', which not only bears small, especially tasty early fruit, but also has exceptional fall color. If your garden is in an area where there are no other persimmons, you will need to also plant males to insure fruit set. Two reliable male cultivars are 'William' and 'George'.

Nothing evokes the warm, lazy feeling of a fall afternoon in the countryside like the sight of two or three persimmon trees lounging against a split-rail fence, their devilishly delicious fruit hanging just out of reach. Truly ripe fruits still on the tree are hard to distinguish from their unfriendly neighbors; a bright orange color does not indicate ripeness. A truly ripe persimmon is extremely soft to the touch and beginning to darken from bright orange. The best fruits are often those that have just fallen from the tree after the first frost. The deer, wild turkey, raccoons, and skunks all know this and will be attracted to a fruiting persimmon tree. Watching the animals is especially fun for children of all ages. Instead of planting the tree near a walk or drive, where the fallen fruit can be a bit of a nuisance, plant persimmons in the naturalized areas of your landscape so that you and the animals can enjoy the persimmon's many delights.

Elaeagnus pungens / Thorny elaeagnus   PLATE 123

Many fragrant woody plants bloom in late spring or summer, but only an iconoclastic few waft their intoxicating perfumes through the autumn landscape. *Elaeagnus pungens*, thorny elaeagnus, is one of those rare few that offer a heady fragrance on the autumn air. A number of *Elaeagnus* species, all with fragrant flowers, are commonly found in the landscape, but thorny elaeagnus stands alone for the sheer intensity of its wonderful spicy scent. The fragrance is deliciously pungent, with lasting overtones of lemon and ginger that hang languidly in the air. The aroma from one plant can easily perfume an entire small garden.

Elaeagnus pungens is a large and gangly evergreen shrub. If left to itself, this plant will rapidly reach incredible proportions in a short period of time—on the order of 15-by-15 feet (4.5-by-4.5 m) in a few seasons—with long streamers of new growth shooting out from it. Judicious, regular pruning each fall to the desired shape and size can tame this tough and durable plant and show off its unique horticultural character, but annual pruning will continue to be required.

The foliage of thorny elaeagnus is quite handsome, with shiny, silvery green leaves on loose, thorny stems. Thorns can get as long as 2 inches (5 cm) but occur relatively sparsely along the stems. The thorns are nasty but inconspicuous in appearance. The leaves have a light, warm brown tint on the undersides. Long stems of these leaves, especially from some of the more interesting cultivars, make wonderful cut material for arrangements. The evergreen foliage is handsome in the winter, particularly if some of the variegated forms are used.

In the fall, tiny, silver-white flowers are borne in the axils of the leaves. The flowers are shaped like a cross between tubular trumpets and diminutive, elongated bells. The flowers may develop into rarely seen red fruits, but the primary attraction of these tiny blooms is their fragrance.

Elaeagnus pungens is such a tough plant that it is considered a weed by some. This native of Japan will grow in almost any site from sun to shade, dry to wet, and it is very salt tolerant. It is completely cold-hardy to zone 6, and it is not subject to any serious pests.

Cultivars of thorny elaeagnus include a number of forms with yellow-gold variegation, such as 'Aurea', 'Dicksonii', 'Golden Rim', and 'Maculata' (or 'Aureo-variegata'). The effect in the landscape of these various cultivars is much the same—bright splashes of gold in the form of a rough shrub. They are certainly not refined, but they are an excellent way to bring bright winter color to a drab, difficult site. Other cultivars include 'Marginata', with silver-white variegation; 'Variegata', with irregular white-yellow variegation; 'Fruitlandii', with larger, more rounded leaves; and 'Simonii', with reduced thorny character and large leaves that are densely silver on the undersides. Surprisingly, thorny elaeagnus can be difficult to propagate from cuttings. Hardwood cuttings are taken in mid- to late winter, after some cold weather. Cuttings taken in early spring, including the previous year's growth, can be rooted successfully under mist. Seed can be sown fresh, but if it does not germinate successfully, then store it at 40°F (4°C) in a moist medium for 3 months before resowing.

Thorny elaeagnus has been used extensively in the Southeast for highway and municipal plantings where it happily grows into huge, unkempt mounds that create an unfair image. While it is true that *Elaeagnus pungens* is not a good choice for small, restrictive sites or formal gardens, it is an excellent choice for larger properties requiring hedging, windbreaks, or cover for a difficult bank. A hedge of thorny elaeagnus is also a perfect refuge for small wildlife, making it a good choice for naturalistic gardeners. In the right garden, the vigor of this zesty shrub is a virtue, and in fall, when its heady fragrance mingles with the scents of fallen leaves and nippy autumn air, thorny elaeagnus is a treasure.

Eleutherococcus sieboldianus 'Variegatus' / Variegated five-leaf aralia
PLATE 124

Each fall, the tones of the garden turn from greens and pastels to golds and reds, purples and browns. At this time of year, it is often refreshing to come across a plant that still offers fresh tones of green and white, even in fall. Variegated five-leaf aralia, *Eleutherococcus sieboldianus* 'Variegatus', offers just this kind of foliage.

In spite of their merits, the use of variegated plants in the garden is often controversial. Variegated foliage, leaves that are of mixed colors, inevitably brings out the stubborn side of gardeners (and we all know just how stubborn gardeners can be). Either variegated foliage is a horticultural gift worth braving civil wars and scaling foreign mountains for, or it is a mutant blight brought into culture by yet another crazed fanatic with no appreciation for the natural world. I have personally witnessed dear friends come near to blows over this issue, which is only topped on the "gardener blood pressure index" by the correlative debate over contorted growth habit.

Setting personal and philosophical arguments aside, however, there are some beautifully variegated plants in the world. Many of these interesting color forms were found as spontaneous changes in the growth of the original plant type—changes in growth that had occurred without any interference from humans—and then were propagated by horticulturists.

One variegated plant that is unarguably handsome is *Eleutherococcus sieboldianus* 'Variegatus', the variegated five-leaf aralia. Until recently, this plant was called *Acanthopanax sieboldianus* 'Variegatus', but taxonomists decided that it was actually a different genus and changed the name. You will still find variegated five-leaf aralia listed in most arboreta and nursery catalogs as *Acanthopanax*, however; it takes years for these changes to finally trickle through all of the plant world. In spite of the frustrations of keeping up with the fickle taxonomists, five-leaf aralia is a good example of why it is important to know botanical names. Variegated five-leaf aralia shares a piece of its common name with members of the genus *Aralia*. While *Eleutherococcus* and *Aralia* are both members of the same botanical family, they are obviously not the same plant. This is a good illustration of why scientific names can be critical for accurate identification of a plant at the nursery, in the garden center, or on a garden plan.

Five-leaf aralia is a tough, handsome native of Japan. It is an upright deciduous shrub whose branches tend to arch outward and over with age. If left to itself, the shrub will eventually become a rather large mass, 10-by-10 feet (3.1-by-3.1 m), in the shape of a broad, upside-down U, but it can be readily pruned to retain the upright character of a younger plant. The tan stems are rough and bumpy, and there are thin but significant thorns at the leaf nodes. Five-leaf aralia flowers in late spring with rounded, branched clusters of tiny whitish flowers, 4–6 inch (10.1–15.2 cm) tall, emerging from the ends of the branches. The flowers mature into shiny black fruit, which is rarely seen in gardens because the flowers are not usually fertilized. The plant is dioecious; primarily only females are grown in cultivation, so males are not present to fertilize flowers.

The leaves are divided into five to seven leaflets with a shape much like the leaves of *Aesculus* spp., the buckeyes, but much smaller, with leaflets only 2 inches (5 cm) long. The foliage of 'Variegatus' is the same size and shape as that of the species, but the color is a refreshing blend of creamy white and emerald-green instead of the species' pretty emerald-green. Some variegated plants show sporadic variegation in different areas of the foliage, but this plant's lively mix of green and white is carried uniformly throughout the whole plant. The amount of white and green are almost equal, a trait that adds to the visual impact in the landscape. The quality of the variegation stays beautiful throughout the growing season—a special plus, because many other variegated plants lose their color, scorch, or become somewhat fungus-ridden during stressful summers. At the North Carolina State University Arboretum, variegated five-leaf aralia adds its wonderfully colored foliage to the mixed shrub border and the white garden.

Variegated five-leaf aralia, like the species, is an extremely adaptable plant. It will perform well in many landscape sites, from dry sand to heavy clay, from full sun to shade, and is completely hardy in zones 4 to 8. It has no serious disease or insect pests. It is especially tolerant of urban settings with extreme conditions and higher pollutant levels. Variegated five-leaf aralia has a somewhat slower, more readily controlled growth rate than the species, which needs tending in very small spaces because of its rapid growth and suckering habit. Pruning, which can be done at any time of year, can keep both species and variegated form in check. Growth may also be slowed down under stress.

Eleutherococcus sieboldianus 'Variegatus' is propagated from softwood cuttings rooted under mist, or by division of the parent plant. The species can also be propagated from seed, which requires a complicated pretreatment involving a sequence of warm followed by cold temperatures. Variegated five-leaf aralia is not common in American nurseries yet, but it is being produced by some specialty nurseries.

The uniformly variegated foliage of *Eleutherococcus sieboldianus* 'Variegatus' makes a dashing addition to a mixed shrub border in sun or shade. It is easy to blend with many different colors and textures in the garden. The shrub could be a refreshing summer specimen of its own, as well as a unique foil for the more fleeting displays of many flowering perennials.

Fraxinus americana / White ash   PLATES 125, 126

The colors of the white ash in autumn are elusive, subtle, and uniquely beautiful in a season otherwise full of flash and brassiness. In early autumn, they display a palette of understated colors rarely seen on other fall foliage—shades of bronze-violet touched with wine and chocolate, or maroon fading to lavender and pale yellow.

Fraxinus americana is a large, deciduous tree that can attain 100 feet (31 m) in height and 75 feet (22.9 m) in width. It has an extensive native range, from the Great Lakes, St. Lawrence Seaway, and Nova Scotia in Canada south throughout most of the Midwest and eastern United States into Texas and Oklahoma. Its shape is upright but becomes more rounded and open as the tree matures. A beautiful shade tree for large lawns and parks, white ash may grow to impressive proportions. The compound leaves are lustrous green on top with pale undersides.

White ash does best on deep, well-drained soils but will also thrive in other soils if they're not too dry. It is hardy to zone 3. Full sun allows for the best quality foliage and growth. Ash trees are unfortunately the favorite of a number of insects and diseases and, in recent years, the trees have been the victims of an "ash decline," as yet not fully understood. If you garden in areas where ash decline has become prevalent, white ash is probably not a good choice for your garden. However, if you garden where there are few *Fraxinus americana,* or if you have existing healthy trees, white ash can be very useful and handsome. In that case, a healthy, vigorously growing tree will not generally be troublesome, and can give many years of beauty in a landscape.

The fruit of any ash tree is a one-winged samara, like a maple's whirligig but with only one wing. Ash trees bear male and female flowers on separate trees, and only the female flowers develop into fruits. Purchasing male trees will prevent you from having to deal with the samaras, which can be a bit of a nuisance near a walkway. Female ash trees are still worth consideration, however, because the fruits, which are born in clusters among the foliage, add a sophisticated note of eye-catching dimension to the trees in late summer.

Many fine vegetatively propagated cultivars of white ash are available in the nursery trade. The cultivars of white ash are generally much more desirable than seedling trees. A few of the many excellent ones include: 'Autumn Applause', known for its maroon fall color, dense branching, and gracefully drooping foliage; 'Autumn Blaze', a female selection with purple fall color; 'Autumn Glory' (Plate 125) offers absolutely stunning fall color; 'Autumn Purple' (Plate 126), a male with excellent displays of purple-red foliage in the fall; and 'Chicago Regal', a vigorous grower that develops purple fall color.

Valuable for its pool of green shade in summer, this fine American tree becomes a strong player in the autumn garden as well. As our eye travels across the fall landscape feasting on rich and brilliant colors, white ash offers subtle refreshment with its blend of antique hues.

Gelsemium rankinii / Swamp jessamine   PLATE 127

What could be better than bright yellow flowers to stave off those end-of-season blues? *Gelsemium rankinii*, the evergreen vine known as swamp jessamine, produces lovely warm yellow flowers that are especially marvelous in the fall.

Swamp jessamine is an almost-identical twin of the more familiar Carolina jessamine (*Gelsemium sempervirens*), which blooms in spring. The only way to tell the two vines apart is the presence of flowers in the fall. This can be confusing in an unusual year when a long dry spell has triggered fall flowering in the Carolina jessamine as well.

Like Carolina jessamine, swamp jessamine will climb just about anything in its path. In spring and again in fall, *Gelsemium rankinii* produces bright lemon-yellow, trumpet-shaped flowers in twos and threes up and down the vine. The flowers are not fragrant, but their warm color is marvelous in the fall. The vines are covered with flowers from March to April and October to November, and a few flowers may open through the winter as well.

Swamp jessamine has the same cultural requirements as Carolina jessamine. It is hardy to zone 7. A native of swampy areas of the Southeast, *Gelsemium rankinii* likes somewhat acid, moist, organic soils. You don't have to have a swamp in your backyard to grow swamp jessamine, though. It is very adaptable and will survive and flower in many different soil types, including heavy clay and sandy loam. The vine flowers best and grows most vigorously in full sun, but it also does quite well in shade although flowers and foliage are somewhat more sparse. There are no significant disease or insect problems. Swamp jessamine is propagated by seed or from hardwood and semi-hardwood cuttings.

Swamp jessamine will twine around itself to form a mound 3–4 feet (0.9–1.2 m) high if grown with no support, but it is even better as a climber on lattice work, rock walls, or fenceposts. The vine is charming when grown up a mailbox post, where the dancing yellow flowers can save a bit of summer for the long evenings ahead.

Ginkgo biloba / Ginkgo, Maidenhair tree   PLATE 128

The *Ginkgo* tree has remained essentially unchanged for over 100 million years. It is a conifer relative but is like no other conifer you can imagine. It is deciduous and its "cones" look like tiny golden persimmons. The leaves are not needles, but flattened, fan-shaped leaves that cover the tree with thousands of emerald-green "fish-tails." In fall, the elegant leaves turn a sparkling yellow that absolutely gleams from a distance.

Ginkgo biloba, ginkgo or maidenhair tree, is the only surviving species in this genus, and it has survived with its character intact from the prehistory of the planet. At one time, it grew in many areas of the world, but today it is "native" to its last remaining outpost in eastern China, where it has been revered and cultivated for the nut inside its foul-smelling fruit. Ginkgo is a tall, deciduous, somewhat slow-growing tree with irregular, horizontal branching. It is usually seen from 20–40 feet (6.2–12.4 m) in the landscape, but ancient trees have reached 100 feet (31 m) or more. It can be extremely long-lived; a tree exists in Korea documented to have been planted 1,100 years ago. The foliage is an unusual divided fan shape with rounded lobes. The leaves are a clear emerald in summer, turning golden in autumn and generally falling together all at once, creating a wonderful shining carpet beneath the tree. The light gray bark is a handsome foil for the fall foliage.

Male and female flowers are on different trees. Female trees bear the malodorous fruit. Encountering the fruit of *Ginkgo biloba* is quite an experience. As it falls from the tree and the soft, fleshy outer covering breaks down, the fruit gives off a very potent aroma that seems to last forever. In addition, trees are usually fairly prolific bearers, and the dropping fruit creates quite a mess. For this reason, male trees are the desirable form. Unfortunately, it is impossible to determine the gender of young seedling ginkgos because the trees do not fruit until they have reached a relatively mature age.

To assure the gender of a purchased tree, buy a grafted cultivar. 'Autumn Gold' is a regular, symmetrical male with uniform fall color; 'Fairmount' is a narrow, upright male form; 'Fastigiata' is a columnar male; and 'Santa Cruz' is an umbrella-form male. 'Pendula' is a generic term used for a number of weeping-type males. 'Saratoga' is a male with rich yellow fall color and a very distinct central leader, which gives it a more formal habit than other selections. 'Variegata' is a male form with golden variegation on the leaves in summer.

Ginkgo biloba will thrive in an incredible range of growing conditions, from extreme cold to heat and humidity to drought. It is very cold-hardy (to zone 3) and transplants readily. It is tolerant of salt and air pollution and has no disease or pest problems. It performs well in a wide range of soil conditions, although it prefers well-drained, sandy, deep soils. In addition to vegetatively propagated cultivars, ginkgo can be propagated by summer softwood cuttings, which are rooted under mist, or from seed, which must be collected in the fall. The pulp is removed and the seed stored in moist sand for 10 weeks at 65°F (16.5°C), followed by three months of cold treatment.

Ginkgo also has another interesting way of propagating itself. The Arnold Arboretum's Dr. Peter Del Tredici, who has studied *Ginkgo* extensively, has observed that trees can form large, rootlike structures near the base of their trunks that grow slowly downwards. Called "basal chichis" (after their resemblance to breasts), these woody growths appear to be "oozing" down the trunk toward the ground. They are especially prevalent on trees found on disturbed, eroded areas and whose trunks have been damaged. Once they grow to the ground, chichis can regenerate both roots, downwardly into the soil, and shoots—effectively regenerating a new tree without producing seed, but one that is contiguous with the "mother" tree through the chichi tissue, and which is a clone of the mother. Ginkgos can develop chichis as young plants in containers as well. You might notice this as a swelling on your plant at or near the soil line. Chichis are one way that *Ginkgo* has insured its survival as a living fossil.

While *Ginkgo biloba* is primarily an ornamental species in the western world, the foliage is also used for medicinal purposes in Europe, as a blood-thinner and to treat circulation-related problems. A unique ginkgo plantation in Sumter, South Carolina, grows more than one million trees to provide leaves for the European market. The trees are grown at very close spacing and the leaves are mechanically stripped off to be processed after harvest.

Ginkgo is not easy to use as a specimen tree in a small landscape because of its size and open, somewhat stark habit; however, its beautiful fall color makes it a special tree for a larger autumn landscape. The trees can often be seen lining avenues and streets, livening parking lots and corporate waste-lands, and surviving the human landscape. Remarkable plantings of ginkgo brighten the Moravian Village in Winston-Salem, North Carolina. Surviving is indeed one of the talents of this oldest of trees, but beauty and color are also an important part of its repertoire. Plant *Ginkgo* in the garden to see this living legacy carpet the ground with its autumn gold.

Hamamelis virginiana, H. macrophylla / Autumn witch hazels   PLATES 129–131

Witch hazels (*Hamamelis* spp.) are a group of deciduous, multistemmed shrubs that bear interesting, ragged flowers of yellow, gold, orange, or red in fall to early spring, depending on the species. Two fall-blooming witch hazels, common witch hazel (*H. virginiana*) and southern witch hazel (*H. macrophylla*), are often neglected by gardeners, although both are excellent shrubs.

Common witch hazel is native throughout most of the eastern United States and crossing the border into Canada around the Great Lakes and Nova Scotia, while southern witch hazel is found from the Carolinas down through Florida. Like the other witch hazels, both of these fall-blooming species have unique flowers about 1 inch (2.5 cm) in diameter with crumpled, very narrow, straplike petals. The flowers are creamy to bright yellow in color and are very fragrant with a sweet, lemony scent. At the North Carolina State University Arboretum, both common and southern witch hazels perfume and color the Piedmont autumn just as the main path exits the white garden.

While both of these are fall-blooming shrubs, they have quite different habits. Common witch hazel, *Hamamelis virginiana*, is a large, open, spreading plant. It reaches as much as 20–30 feet (6.2–9 m) in the wild but is usually seen at closer to 15 feet (4.5 m) in a landscape setting, with an equal spread. Reliable to zone 4, *H. virginiana* is the most cold-hardy of the witch hazels. Wild plants are often quite disheveled in appearance, but when grown in full sun in cultivated areas, the shrubs are tidier. In the landscape, the coarse, irregular branching becomes a bonus for winter interest. In the fall, the glossy green foliage turns bright golden yellow (Plate 129). This sometimes occurs just as the flowers open, making it difficult to see their canary-yellow display. But their fragrance still permeates the air, and often the leaves fall while the flowers are still showy. A medicinal extract is made from distillations of the bark from young stems and roots of common witch hazel.

Southern witch hazel, *H. macrophylla*, is a much more demure shrub than common witch hazel. It reaches 6–10 feet (1.8–3.1 m) with a spread of about 5 feet (1.5 m). It is hardy to zone 6. Branching is finer and more dense than that of its larger cousin. The flowers are a light, creamy yellow, smaller and not quite as showy as those of common witch hazel. The soft green foliage is also somewhat smaller (in spite of its specific epithet, *macrophylla*, which translates to "large leaves"), with furry undersides. Fall color is a lovely blend of rosy purple, bronze, and gold.

Witch hazels prefer moist soils and will grow well in full sun or partial shade. They are tough plants that will tolerate urban conditions as long as they are not too dry. There are no serious diseases or pests to contend with. Roots are sensitive to disturbance; move plants either in containers or balled-and-burlapped to avoid problems following transplanting.

Vegetative propagation of these witch hazels is difficult and is probably best left to nursery professionals and serious horticulturists. Cuttings of very soft, succulent young growth are wounded slightly by scoring or scraping the cut end, then dipped in rooting promoter and rooted under mist. They may take in the range of three months to root, and a large percentage of the rooted cuttings will die following rooting. Seed requires two resting periods at different temperatures before it will germinate, but seedlings are often abundant under landscape plants.

Fall-blooming witch hazels are underutilized in most gardens, even though they contribute both perfume and color to the autumn landscape. Their handsome fall foliage combines with appealing fragrance and extraordinary flowers for a distinctive display. They are very effective when used in a mass and are excellent when planted in mixed groupings of other plants with red and purple fall color. Both mass and single specimens also make striking foils for fall-blooming perennials, such as purple-flowered asters and burgundy or red-flowered chrysanthemums.

Heptacodium miconioides / Seven-son flower   PLATE 132

Heptacodium miconioides, seven-son flower, is a rare and magnificent deciduous shrub that offers unique landscape character every month of the year with its foliage, flowers, fruit, and peeling bark. The creamy white, sweetly fragrant flowers cluster in pyramidal tiers at the ends of the branches throughout the fall. The Chinese name for the plant, "seven-son flower from Zhejiang," refers to the number of individual flowers in each flower cluster. Even the botanical name *Heptacodium* makes reference to this character, with *hepta* meaning "seven" and *codium* referring to a "head." The flowers are unique among those of other woody plants because the buds are formed in spring and slowly develop over a five-month period, finally blooming in autumn.

But while the flowers are quite lovely (and especially appealing since they are showy at a time when few other woody plants are in bloom), they are not the most dramatic trait of *Heptacodium miconioides*. In late fall, the flowers mature to small, rounded fruits, each of which is crowned with a persistent calyx that turns a bright cherry-red or rosy purple. The effect of all the tiers of these exotic fruits with seven splendid purple crowns is spectacular.

The foliage of seven-son flower is beautiful in its own right. Large, glossy green, narrowly heart-shaped leaves emerge in spring and quickly develop three very deep veins running the length of each leaf. Leaves remain unblemished and handsome late into the fall, when they finally drop without developing significant fall color.

Even in the dead of winter, when neither flowers, fruit, nor foliage are present, seven-son flower offers fascination for the garden. The light tan bark peels away in thin, curling strips to reveal a nutty red-brown underneath. The trunks of the shrub look as if thick beige netting had been wrapped around its cinnamon stems. The warm color is welcome in the winter landscape.

Heptacodium miconioides eventually reaches about 15–20 feet (4.5–6.2 m) in height, with a spread of 8 to 10 feet (2.4–3.1 m), in an open, fountainesque shape. A large, multistemmed shrub by nature, the plant can be pruned to create a single- or several-stemmed small tree, perfect for small-scale modern landscapes.

Seven-son flower thrives in a range of soils from dry sand to wet, heavy clays. It is cold-hardy to zone 5 and has no significant disease or insect problems. It flowers and fruits most prolifically in full sun but will remain attractive in partial shade. (Dense shade results in a leafy, leggy plant.) It is easily propagated from stem cuttings taken in spring and summer and rooted under mist. It can also be propagated from seed, but seed requires both cold treatment and scarification.

A relative of honeysuckle (*Lonicera* spp.) and forsythia (*Forsythia* spp.), *Heptacodium miconioides* was first discovered in China by the famous plant explorer, E. H. Wilson, and finally brought to cultivation in the West at the Arnold Arboretum. The plant was originally distributed in the West with the botanical name of *Heptacodium jasminoides* (and is still listed by that name in many catalogs and plant lists). While modern gardeners will not have to search as hard as Wilson did to locate seven-son flower, they will still find it a challenge to procure this unusual plant. A number of specialty mail-order nurseries are working to make it more available.

Seven-son flower makes an excellent specimen plant. It can also be used in a mixed border, or combined with other, lower-growing shrubs in a massed group. At the North Carolina State University Arboretum, seven-son flower keeps company with the Colorado blue spruces (*Picea pungens*), in the east side of the gardens. Watching the crown of this unique plant transform itself from floral cream to berried, bright rose is a special fascination in the fall garden.

Ilex 'Mary Nell' / 'Mary Nell' holly   PLATE 133

Ilex 'Mary Nell', a relatively new selection from the famous Tom Dodd Nursery in Alabama, is a lush beauty with three hybrid parents: *I. cornuta*, the tough, heat-tolerant Chinese holly; *I. pernyi*, Perny holly, a common breeding parent with neat, small foliage; and *I. latifolia*, lusterleaf holly, a large-leaved holly with unique broad, blue-green leaves. From these three parents, 'Mary Nell' has inherited tough landscape adaptability; large, glossy, dark blue-green leaves; a dense, pyramidal to columnar shape; and the tendency to produce prodigious quantities of bright red fruit. In fact, 'Mary Nell' may well outpace 'Nellie Stevens' as a top evergreen holly for gardens with poorly drained soils in hot climates.

'Mary Nell' will eventually reach 20–25 feet (6.2–7.5 m) in height with a moderately formal pyramidal shape and very dense foliage. It grows well in partial shade, but foliage quality and fruiting will be best in full sun. 'Mary Nell' will take almost any soil a gardener can dish up and thrive beautifully. It is hardy to zone 7 and may suffer foliar burn and twig kill in severe winters. Like most evergreen hollies, 'Mary Nell' is easily propagated from hardwood or semihardwood cuttings rooted under mist.

'Mary Nell' is the perfect choice for a dark, rich hedge or foundation planting, adding depth and beauty. It also makes a dramatic corner plant, and its formal shape can be used in a large container on a spacious patio or courtyard. The broad, dark foliage makes a perfect foil for winter-flowering and variegated plants. The plentiful fruit are a cheerful addition to the winter garden and in cut holiday arrangements, and they are relished by hungry birds as well.

Liriodendron tulipifera / Tulip tree   PLATE 134

The bright canary-gold fall foliage of the native tulip tree, *Liriodendron tulipifera*, is one of the boldest elements in the fall landscape in the eastern half of the United States and in a few areas of Canada at the border near the Great Lakes. Tulip tree can be found growing in mixed hardwood forests on a range of soils. It is one of the tallest native trees, with the potential to reach over 100 feet (31 m) in the wild.

In the wild, tulip trees are usually very tall, with perfectly straight, handsomely etched, light gray trunks that do not branch until near where the crown emerges from the forest canopy. The upper, branched part of wild trees creates a rounded oval crown in the forest canopy. These very straight trunks have been traditionally prized for lumbering purposes. Although it is also known as tulip poplar or yellow poplar, this magnificent large tree is not actually a true poplar (*Populus* spp.). It bears a superficial resemblance to poplars in having a generally tall, upright habit, but it is actually a close relative of *Magnolia*.

Liriodendron tulipifera develops a somewhat different habit if grown in an open garden than it does in the more restrictive environment of the woods. In the landscape, young trees are branched much lower down on the trunk, creating a more pyramidal outline. As trees mature in an open landscape, and their crowns become taller, the trees assume a more upright, oval shape, but they retain a more rounded, branched crown than most trees in the woods develop. Established landscape tulip trees are generally seen in the 40–80 feet (12.4–24.8 m) range with a spread of 20–50 feet (6.2–15.2 m). Whether grown in the wild or the garden, all tulip trees become bright golden pools in autumn.

Autumn is the season when tulip tree is the most showy, but it also has exceptional character during the other three seasons of the year. In spring and summer, the tree holds large, velvety, bright green leaves, 6–8 inches (15.2–20.3 cm) long, and shaped like flattened, two-thumbed mittens—distinctive even from a distance.

In spring, after leaves have emerged, one of nature's most amazing flowers is displayed in the tops of the tulip trees. The large blooms are tuliplike in shape, but they are an unbelievable chartreuse color, with petals splashed at the bases with a neon orange that is breathtaking up close. Tulip trees generally take five to eight years to flower, which means the flowers are borne only after the tree is relatively tall. Unfortunately, the flowers are generally borne on the tops of the uppermost branches, facing the sky, and so are rarely available for close inspection in the wild until they drop. The spectacular blooms can frequently be seen at closer proximity on garden-grown trees or trees in open sites, although the vivid flowers are surprisingly well camouflaged against the foliage. The flowers are also worth seeking out for their unique, elusive scent. The flowers mature into an interesting conelike fruit, similar to that of its close relative Southern magnolia (*Magnolia grandiflora*), but with tan, papery seeds instead of the magnolia's red, berrylike seeds. These tan "cones" on tulip tree persist, upright on the branches, for several months, adding interest to the winter landscape.

Tulip tree prefers moist, well-drained, loamy soils in full sun, but it will tolerate heavier clays and is quite rapid in growth rate, as much as 3–4 feet (0.9–1.2 m) in a year. It is completely hardy to zone 4 but growth will be slower in the northerly parts of its range. When stressed, *Liriodendron tulipifera* is subject to a few problems like aphids and the associated black sooty-mold. It is very important to keep tulip tree regularly watered during droughty periods to avoid premature yellowing and drop of the leaves. With good horticulture, none of these should be serious problems and tulip tree will grow into an exceptional large tree. Propagation is generally by seed, which requires a moist, cold pretreatment of three months before sowing. The cultivars are propagated by grafting.

A number of good cultivars with unique traits are available from specialty nurseries. 'Arnold', a repropagation of the cultivar 'Fastigiatum', was named for the Arnold Arboretum, where the scion wood was collected, and was put into production by Monrovia Nursery in California. 'Arnold' is a fastigiate, columnar form reaching 50–70 feet (15.2–21.7 m) with a spread of only 15–20 feet (4.5–6.2 m). It would make an extraordinary tall, columnar hedge along a park drive, and would be fabulous combined with a tall, dark evergreen in a double allée planting—imagine it in full fall color in October. 'Fastigiatum' is the original narrow, columnar form. 'Aureo-marginatum' has striking lime-yellow margins on the leaves. 'Compactum' is a dwarf form from J. C. McDaniel in Illinois, with excellent potential for smaller landscapes. 'Integrifolium' is missing the "thumbs" from its mitten-shaped leaves, so that the leaves are shaped like a box. 'Medio-pictum' has a lime-yellow center on its leaves. There are also some contorted forms (with rather confused nomenclature at this point): 'Contortum', 'Crispum', and 'Tortuosum'.

The blaze of color that is tulip tree in the autumn is just as stunning in landscapes and gardens as it is in the fall woods. *Liriodendron tulipifera* is an excellent choice for parks and large properties where the roots can spread and it can reach its grand height. The long allées of mature tulip trees planted 90 years ago at the Biltmore Estate in Asheville, North Carolina, show the potential majesty of this tree.

Nyssa sylvatica / Black gum   PLATES 135, 136

The foliage of *Nyssa sylvatica*, black gum, sparks its native woods like the glow of a dozen scarlet blazes, standing apart brilliantly from other trees. Native to the eastern half of the country and across the border into Canada near the Great Lakes, black gum can be seen growing in a variety of sites, from abandoned fields and dry ridges to cold, wet swamps (*Nyssa* means "water nymph").

Black gum, also known as black tupelo, is a large, slow-growing tree that can reach heights of 100 feet (31 m) but is usually seen at 20–40 feet (6.2–12.4 m) in the landscape. It is a beautiful, upright tree with dense branching. Young black gum trees have a somewhat oval or pyramidal shape that eventually matures into a spreading, rounded crown. The lower branches of young trees often have a slightly pendent habit (similar to that of young pin oaks, *Quercus palustris*).

The foliage of black gum is a handsome lustrous green through the spring and summer. The tiny white spring flowers and little gray-blue fruits, which develop in early fall, are not especially showy. Autumn is when black gum is at its peak. Fall color on individual trees ranges from intense crimson red to bright canary yellow and orange. Whatever autumn shades any given tree develops, they are certain to be dramatic and dependable every year. Another attractive feature of black gum is its dark charcoal bark, which develops a broken, blocky texture like alligator hide.

Nyssa sylvatica has a somewhat undeserved reputation for being impossible to transplant. This reputation comes from unsuccessful efforts of moving wild trees from forested areas, a difficult and generally futile process. Black gum has a taproot and does require careful transplanting, but with a little care, it can be successfully planted in many sites. Small, balled-and-burlapped nursery-grown trees are the best choice for new plantings.

Black gum prefers moist, acid, well-drained soils, but it also grows well in heavy clay soils. It will grow in full sun or partial shade. Full sun will result in the best fall color. The tree has no serious disease or pest problems and is hardy to zone 3. It is readily propagated from seed. Germination is enhanced by storage in a moist medium, such as peat moss, for two to three months at 40°F (4°C), before sowing.

Nyssa sylvatica 'Dirr's Selection' (Plate 135) is an unnamed selection by plantsman Dr. Michael Dirr made for its brilliant fall color. Because vegetative propagation of *Nyssa sylvatica* is very difficult, however, there are no readily available cultivars of black gum.

Three other species of *Nyssa* offer unique attributes for the garden. *Nyssa aquatica*, water tupelo, is a southern swamp tree which grows in areas with periodic flooding, but it will thrive in drier sites as well. This tree's trunk is somewhat swollen at the base, then narrows and tapers distinctly. Water tupelo is also a large tree (to 120 feet [38 m]) somewhat more open in habit than black gum, and it has noticeably larger leaves, to 7 inches (17.7 cm), and larger, purplish fruit. It is less cold-hardy than *N. sylvatica*, being reliable to zone 6.

Nyssa ogeche, Ogeechee tupelo, is native to the southeastern corner of the United States (South Carolina, Florida, Georgia). It has lighter gray bark and lighter green leaves than black gum, and is a smaller tree, reaching 30–50 feet (9–15.2 m). The undersides of the leaves are silvery. Ogeechee tupelo bears red, edible fruits that have been used as a citrus substitute. It will thrive in diverse conditions, including wet and swampy soils and well-drained loam. The tree does not develop showy fall foliage color. It is reliably hardy to zone 7.

Nyssa sinensis, Chinese tupelo, is quite rare and very handsome. Branching in this native of China is a bit less dense and more uniform than black gum, but fall color is equally attractive. It is hardy through zone 7 and possibly into zone 6.

Black gum is an attractive tree as a specimen or in a naturalized setting throughout the year. As autumn takes hold of the woods and gardens around us, the brilliant hues of black gum foliage will shine out from all the other colors. The colorful trees are a reminder of the beauty that can result, year after year, from planting just one special tree.

Osmanthus fragrans / Osmanthus   PLATE 137

One fall evening, on a stroll through the garden, the air will be perfumed with an incredible heady aroma. The scents of oranges, cinnamon, apple blossoms, and many more fragrances will hover deliciously around the trees and hedges, drawing you along every curving path in search of the source of this treat. If you are patient enough, you will eventually discover that this wonderful aroma comes from the tiny flowers of fragrant tea olive, a broadleaved evergreen shrub with wildly fragrant flowers and quietly handsome foliage.

Osmanthus fragrans is a large shrub or small, multistemmed tree. In general, *Osmanthus* are similar in appearance to hollies (*Ilex* spp.), so much so that the common name for one group of *Osmanthus* is false holly. They are easily distinguished, however, because the leaves of true hollies are arranged alternately along the stem while those of *Osmanthus* are oppositely arranged. Native to Asia, *Osmanthus fragrans* is exceptional for its beautiful evergreen foliage and fabulously fragrant flowers. The rich green, glossy leaves are 2–4 inches (5–10.1 cm) long and half as broad. The foliage retains its quality throughout the winter. In midautumn, tiny flowers open in the axils of the leaves. The flowers are greenish-white and very inconspicuous (with one important exception described below) and create a cloud of perfume for several weeks. This plant can reach 20–30 feet (6.2–9 m) and can be limbed up to create a dramatic, multistemmed tree or pruned to maintain a more conservative stature for smaller landscapes. The smooth, gray bark is also attractive.

Fragrant tea olive prefers well-drained, moist, acid soil, but it will perform well in many landscape conditions. It is quite tolerant of heavy clay soils and will take full sun and many of the stressful conditions of urban areas. Hardy through zone 8 and the milder parts of zone 7, fragrant tea olive is the least hardy *Osmanthus*; it will suffer foliar injury at temperatures less than 0°F (−16°C). After a severe winter, drastic pruning or replacement may be needed because of cold damage. *Osmanthus fragrans* is propagated from seed or cuttings. Newly hardened early summer cuttings are treated with rooting promoter and rooted under mist.

Osmanthus fragrans f. *aurantiacus* is a spectacular color form, sometimes referred to as the cultivar 'Aurantiacus'. This amazing plant bears light sherbet-orange flowers in prolific clusters each fall. While the flowers are as tiny as those of other fragrant tea olives, their numbers and color make them especially showy against the deep green foliage. One plant can perfume an entire acre with its delicious scent. Cut flowering stems are beautiful in arrangements and will fill even large rooms with their aroma. *Osmanthus fragrans* f. *aurantiacus* is an excellent specimen plant for fall and winter interest in almost any garden, planted where it will be discovered on a garden wander, or near a patio or seating area. This form also makes a special container plant, an arrangement that allows the gardener to move the plant from place to place when in flower.

A number of other species and cultivars are suited to gardens. Almost all of these flower in the fall and are fragrant to one degree or another. It is important to be clear about which _Osmanthus_ you are interested in when purchasing plants at nurseries and garden centers.

Osmanthus heterophyllus, false holly, generally has foliage very similar to many evergreen hollies and is a common landscape foundation plant. Hardy to zone 7, it is distinctly different in foliage type from _Osmanthus fragrans_. Many cultivars of this plant, ranging from spiny-leaved holly look-alikes to variegated forms, are commonly available. 'Rotundifolius' has small, smoothly rounded, leathery leaves and looks very different from any other _Osmanthus_. It does have the fall display of fragrant flowers.

Osmanthus × fortunei, Fortune's osmanthus, is a hybrid of false holly and fragrant tea olive and can be confused with either because it is intermediate between the two in appearance. Fortune's osmanthus bears intensely fragrant flowers in mid-autumn. One very commonly seen form of this hybrid is the attractive cultivar 'San Jose', with larger foliage than the parent hybrid.

Osmanthus americanus, devilwood, is a native southeastern United States plant with larger foliage and a somewhat more coarse and open habit than the other _Osmanthus_. Devilwood is extremely hardy and should take winter temperatures as low as −18 to −20°F (−25 to −26°C). The leaves are dark olive green and leathery, almost like some of the _Viburnum_ species, and very striking. The bark is somewhat lighter than other _Osmanthus_. Individual flowers are very similar in appearance and perfume to fragrant tea olive, but they bloom in spring instead of fall and they are borne in upright clusters in the upper branches of the plant instead of in the leaf axils up and down the stem. This is an excellent and underutilized plant that is available from only a few specialized nurseries because it is extremely difficult to propagate.

Osmanthus fragrans and its wonderfully scented relatives are a wonderful way to bring the beautiful foliage and fragrance of _Osmanthus_ to the landscape. Their rare combination of elegant evergreen character and dazzling fragrance make extraordinary hedges, masses, and foundation plantings. Osmanthus offers foliage and fragrance to enhance the atmosphere of any garden.

Oxydendrum arboreum / Sourwood   PLATE 138

Our native sourwood, _Oxydendrum arboreum_, flaunts its rich fall color far ahead of others, standing out like a colorful portrait in a room full of black and white still-lifes. Called sourwood because of the sour taste of its sap (even the Greek "oxy-dendrum" translates to "sour tree"), this tree is a close relative of rhododendrons, azaleas, and mountain laurels. In autumn, the leaves turn wine-red, pink, and lavender, often mixed with some yellow shades as well. This watercolor foliage sets off the even more delicate ornament of persistent, creamy tan fruit clusters set among the fall colors. But fall is not this tree's only star season.

Sourwood is native to the middle and southeastern United States and can be found growing on the edges of woods, along roadsides and bluffs, and on the moist slopes of hardwood forests. It is most common in the mountain and upland regions but is also found more sparsely scattered throughout Piedmont and coastal plain areas.

Oxydendrum arboreum is a small tree that grows quite slowly. It generally reaches

15–20 feet (4.5–6.2 m) in a landscape, but taller specimens can sometimes be found. The branches are gently pendulous, creating a graceful silhouette. Foliage is a glossy emerald-green in the spring and becomes rich dark green in summer.

In summer, sprays of tiny white flowers dangle in clusters among the leaves. Many sourwoods flower so heavily that their foliage is barely visible. From a distance, they appear to be covered in silvery white lace. Honeybees love sourwood flowers and make delicious sourwood honey from the nectar of these tiny blossoms. Sourwood is sometimes called "lily-of-the-valley tree," because of the resemblance of its flower clusters to those of the familiar old-fashioned garden flower grown from rhizomes. After leaf-fall, sourwood's refined branching habit and handsome, dark gray bark are artistic additions to the winter landscape. The bark is often furrowed into interesting blocks, much like the bark of persimmons.

Culture of sourwood is similar to its *Rhododendron* relatives. It prefers well-drained, moist, acid soils high in organic matter. Sourwood can also be grown in somewhat more drier soils than *Rhododendron*, but it suffers in heavy clay and may die or fail to thrive under those conditions in the landscape. The tree flowers most profusely and gives the best fall displays in full sun, but it will perform well in partial shade also. It has no serious disease or pest problems and is hardy to zone 5. It is propagated from seed germinated under mist.

Seedling trees vary greatly in terms of flower profusion, fall color, and even growth habit. This is obvious among wild trees seen along roadsides and in the woods. Unfortunately, no cultivars of sourwood are readily available. Specialty nurseries occasionally offer contorted or prostrate forms, but this is very rare. Sourwood is essentially impossible to propagate from cuttings, which explains the lack of cultivar development. It can, however, be successfully reproduced by grafting or tissue culture, and we can hope that propagation of cultivars named from superior seedling selections will follow. Certainly this special tree deserves more horticultural attention than it has received to date.

Oxydendrum arboreum makes a spectacular specimen tree in almost any garden. Sourwood's year-round interest—lovely flowers, fall color, and graceful form—bring understated sophistication to a landscape. As fall begins to cloak the woods and fields in brilliant colors, look for the special beauty of sourwood, a beauty that belongs in the garden as well as the woods.

Parrotia persica / Persian parrotia   PLATES 139, 140

Trees and shrubs that develop or keep their fall color late into the season seem especially showy against the gray and brown etchings of the other already-winterized plants. One such plant that leaves us with a bright autumn encore is *Parrotia persica*, Persian parrotia.

Native to Iran (which was once called Persia), this tree's genus name was given in memory of F. W. Parrot, a German naturalist, although the natural association of this tree's name with a brightly colored bird is certainly appropriate in the fall.

Persian parrotia is a small to medium-sized deciduous tree with somewhat upright branching. In cultivation, it reaches 15–30 feet (4.5–9 m) in height (larger, older specimens may be found in arboreta and botanical gardens) with an equal to greater spread, which creates a broadly rounded to oval habit. The bark of a young tree is a smooth, warm gray reminiscent of the bark of beeches (*Fagus* spp.). As parrotia matures, the bark peels to create a puzzle of silver, green, and cinnamon pieces (Plate 139), somewhat like the bark of older lacebark pine (*Pinus bungeana*). The flowers and fruit of parrotia are not very showy, although the scarlet stamens of the flowers are bright and interesting upon close inspection.

The foliage emerges tinged with deep burgundy. It matures to a glossy emerald-green in the summer and then bursts into a spectacular fall display of yellow-gold, pumpkin-orange, and crimson. Parrotia foliage is very similar in size and shape to that of its close relatives witch hazel (*Hamamelis* spp.) and fothergilla (*Fothergilla* spp.). During its fall display, garden visitors are invariably stumped as to its identity.

Parrotia persica is completely hardy to zone 4 and has no pest or disease problems. It prefers well-drained, loamy soil in full sun but also does very well in heavy clay and in light shade. Parrotia should be moved balled-and-burlapped. It may be a bit slow to establish but adapts well once in place.

Parrotia is propagated from seed or cuttings. Seed propagation is relatively complex, requiring both warm and cool treatments for several months. Propagation by cuttings is simpler, with cuttings taken in early summer, treated with rooting promoter, and rooted under mist. Rooted cuttings are left in the rooting medium through one winter period and are potted when the next season's new growth has begun to emerge.

There are no cultivars of *Parrotia persica* available, except for one exceedingly rare weeping form, but the seedlings and unnamed clones make beautiful trees in the landscape. Each autumn at the North Carolina State University Arboretum, *Parrotia* can be found bringing Persian gold to both the east and west gardens of the Arboretum.

Parrotia is a show-stopper in the fall, but it is also elegantly handsome in all the other seasons of the year. *Parrotia persica* is a perfect choice for a specimen tree in small gardens, or for a tough street tree in somewhat restricted spaces. It is a dynamic partner for dwarf conifers and broadleaved evergreens, and a beautiful companion for perennials and flowering shrubs in mixed borders. A mature tree with exfoliating bark is a real horticultural jewel.

Photinia villosa / Oriental photinia   PLATES 141, 142

Redtip was once a tough landscape plant with predictable, showy red new foliage, but the ravages of disease and overuse have limited its effectiveness. Because of its ubiquitous nature, *Photinia* × *fraseri*, common redtip, has dragged the entire genus of *Photinia* with it on its own descent into ill repute. One exceptional *Photinia* that deserves greater attention and use than it currently receives is Oriental photinia, *P. villosa*. Oriental photinia is strikingly different from its common, maligned cousin, and it really shines in the fall, when its leaf color changes to handsome combinations of bronzy gold and dark red and bunches of 0.5-inch (1.2-cm) scarlet fruit, resembling hawthorn berries, hang from the branches. The berries are so brilliantly colored that they appear painted with a startling, fire-engine-red that glows brightly from the foliage.

Photinia villosa is a large, deciduous shrub or small tree, reaching 10–15 feet (3.1–4.5 m) in height with a spread of 8–12 feet (2.4–3.7 m). The habit is somewhat like a broad, informal umbrella. The wide, teardrop-shaped leaves are crowded almost into bunches on the stems. The foliage is a bright, warm green and retains good quality throughout the summer heat, changing to autumnal tones of bronze and red in fall and winter. In late spring, small white flowers are borne in lacy clusters. The delicate flowers are not showy, but they are responsible for the vivid red berries that decorate the plant in fall.

Oriental photinia is native to China, Japan, and Korea. Like many other Asian natives, it is a dependable plant in a range of conditions, including areas with hot, wet summers and heavy clay soils. It is reliably hardy to zone 4. *Photinia villosa* does best in full sun but will also grow well in light shade. Fireblight disease can be a problem in areas where there are other infected plants nearby. Oriental photinia prefers acidic, moist, well-drained soil but also performs well in clay. In spite of the fact that common redtip is generally sheared and pruned to induce unending crops of shiny new, raw-red foliage, none of the photinias actually require pruning to develop good shape and vigorous growth.

Photinia villosa is propagated from seed, after a two-month pretreatment of cold temperatures. Oriental photinia is a challenge to propagate from cuttings, but it can be rooted from softwood cuttings harvested in early summer, treated with rooting promoters, and rooted under mist. The North Carolina State University Arboretum at Raleigh, North Carolina, has selected and propagated a superior cultivar, 'Village Shade' (Plates 141, 142), which is entering production and will be available in the nursery trade in the future.

Gardeners have a propensity to latch onto single plants as the one solution to an array of landscape designs and challenges. These types of horticultural obsessions, focusing on plants such as common redtip and Bradford pear (*Pyrus calleryana* 'Bradford'), set us up for epidemic disasters both in terms of plant disease and gardener disease. Diversity in our choices of plants for gardens and landscapes offers an opportunity to create environmentally sound gardens using an exciting and infinite garden palette. Oriental photinia is a good start for diversifying our shrub plantings. It makes a fine specimen plant for small gardens and is a lovely addition to a mixed border, especially for fall and winter character.

Pistacia chinensis / Chinese pistache tree   PLATE 143

Pistacia chinensis, Chinese pistache, is a deciduous, long-lived tree that flames into color every fall with colors to beat those of many of the best sugar maples. The foliage turns a deep, brilliant red and shades to orange as the season progresses, even in warm climates. A good Chinese pistache tree can be relied on to put on a wonderful show in the fall of each year—even in climates where hot weather persists into fall, preventing good fall color development on other trees.

Pistacia chinensis reaches 30–40 feet (9–12.4 m) in height and spreads to 30 feet (9 m). The tree grows to have a dense, oval crown, which makes it an excellent shade tree. Female trees have a more irregular crown shape than the males. The foliage is divided, like that of a pecan tree, and the leaves are a glossy green in summer. Chinese pistache is dioecious; that is, it has male and female flowers on separate plants. When planted in the vicinity of male plants, the females produce attractive, bright red fruits that mature to a soft blue in early fall (when the birds don't devour them, that is).

Chinese pistache was introduced from China, where the young leaves and shoots are eaten as a vegetable. In California, this tree has been used as the understock for the edible pistachio tree, *Pistacia vera*, but the fruits of the Chinese pistache itself are not edible. As a landscape tree, Chinese pistache does well in urban growing conditions, including areas with restricted space for root growth and in conditions of drought or low fertility. It is cold-hardy to at least zone 6, but wild populations in China may be sources of seed of more cold-hardy genotypes. Chinese pistache harbors no troublesome diseases or pests and is very tolerant of drought and poor soils.

The most striking character of *Pistacia chinensis* is its fall color; however, there is great variability in shade and intensity of color from tree to tree because the plant is propagated from stratified seed, due to the difficulties of asexual reproduction.

The softly textured foliage of the Chinese pistache makes it a beautiful addition to the landscape for summer's mantle of green as well as autumn's crimson hues. Because it is so tolerant of less than perfect conditions, *Pistacia chinensis* makes an excellent street tree. It is also a good choice for a shade tree for suburban yards and gardens, where growing conditions are frequently less than ideal. This special ornamental tree can add a dash of dramatic color to almost any landscape.

Prunus subhirtella 'Autumnalis' / Autumn-blooming Higan cherry
PLATES 144, 145

The amazing *Prunus subhirtella* 'Autumnalis' flowers delicately throughout the fall and winter, almost as if it is warming up for its more intense display in spring. Deep rose buds open to soft pink blossoms which turn almost white before fading. The fall and winter flowers open sporadically throughout the canopy, as if a subtle designer had strategically placed just the right touch of color against a dark frame. Spring brings a more exuberant display, in tune with all the other flowering trees in the spring garden, but the special quality of this tree lies in finding its branches decorated in pink above the sparkle of a light winter snow.

A number of cultivars and botanical varieties of the species *Prunus subhirtella* are popular garden ornamentals, but 'Autumnalis' is distinctive for its upright branching habit and its single to semidouble light pink flowers that open almost year-round. The tree will reach 20–40 feet (6.2–12.4 m) in height with a rounded crown spreading to 15–20 feet (4.5–6.2 m). The foliage is an attractive light green in the classic ovate cherry-leaf shape. The gray-brown bark is speckled with the prominent lenticels of most cherries (although this characteristic is not as pronounced as on some other *Prunus*).

Like other *Prunus subhirtella* varieties and cultivars, 'Autumnalis' is very stress tolerant. Hardy to zone 4, it will perform well under extremes of both heat and cold, making it an excellent plant for much of the United States and milder areas of Canada. It grows beautifully through heat, drought, ice storms, and cold winters. Many cherries are considered difficult in the landscape, but this selection is quite tough and relatively disease- and insect-free. It will thrive in both dry soils and moderately heavy clays and does best in full sun. Like most cherries, it requires good care (including regular watering) to become established, but once settled in, it is very tough and care-free. 'Autumnalis' does best in full sun but will also tolerate partial shade with some reduction in flowering. It is propagated from semihardwood, leafy cuttings taken in early summer and rooted under mist.

Prunus subhirtella 'Autumnalis' brings fresh pink magic to the fall and winter garden. It is especially useful as a small tree along winding paths and as a focal point in mixed plantings. It makes a delightful addition to a patio or walled garden and is lovely over fall-blooming bulbs, such as autumn crocus. The late blossoms of this tree are a delightful surprise for the landscape's quiet seasons. Because of their delicate nature, the flowers never jar the eye from its early winter sensibilities, but instead add an enchanting pastel charm to the season's more somber hues.

Pseudolarix amabilis / Golden larch

PLATES 146, 147

Golden larch is a dramatic deciduous, coniferous tree. Like bald cypress (*Taxodium distichum*) and dawn redwood (*Metasequoia glyptostroboides*), golden larch drops its needles in the fall and remains bare-branched all winter until the new leaves emerge in the spring. What is wonderful about golden larch is the spectacular golden color which the leaves develop shortly before they fall, creating an incredible golden pyramid.

While its descriptive common name declares this large tree a larch, it is not a true larch (*Larix* spp.), but only a close botanical relative. The previous taxonomic incarnation of golden larch was *Pseudolarix kaempferi*.

Golden larch has lovely foliage throughout the growing season. In the spring, the needles emerge a soft emerald-green and then develop a bluish bloom through the summer. The needles are arranged in whorls along the branches. In the fall, the needles turn a rich red-gold. The fall color of golden larch foliage is quite amazing, but it is relatively short-lived. The needles generally remain on the tree in full fall regalia for only ten days or so. The unusual-looking cones of golden larch are also beautiful, ripening from the blue-green of summer to light golden brown in autumn, but they are usually found near the top of the tree, which makes it difficult to appreciate their display.

Golden larch is native to the mountains of eastern China, where it grows to over 100 feet (31 m) tall. In garden settings, it is usually seen at heights of 20–40 feet (6.2–12.4 m). The tree has open, spreading branching that develops quite a broad spread for a conifer as it matures. The habit of older specimens is somewhat reminiscent of that of cedars (*Cedrus* spp.). Because of its growth rate of 1 foot (0.3 m) or less a year in cold areas, *Pseudolarix amabilis* has a reputation for being a slow-growing tree, but the specimen at the North Carolina State University Arboretum in Raleigh, North Carolina, has grown about 4 feet (1.2 m) a year.

Golden larch is completely hardy to zone 4. The tree has no serious disease or pest problems. It prefers light, moist, acidic, and well-drained soils, but it is tolerant of heavy clay soil. Golden larch does best in full sun and appears tolerant of polluted urban air.

Pseudolarix amabilis is propagated from seed, but it is difficult to propagate. The seed is rarely fertile from single specimen trees; it requires cross-pollination for the production of viable seed. If two or more seedling trees are planted together, their seed will likely be fertile. Seeds give the best germination when held under moist conditions at 40°F (4°C) for two months before sowing.

Golden larch is a relatively rare tree. It is available from some specialty nurseries. 'Annesleyana' is an extremely rare golden larch cultivar. It is a semidwarf, densely bushy plant with somewhat more pendulous branching than the species.

There are few sights that can match a golden larch at the peak of its fall color. Some of the oldest and most beautiful specimens can be seen at the Arnold Arboretum, Jamaica Plain, Massachusetts. This magnificent tree is one of the most unique specimens for the garden that you could wish for. It can be readily used as a special interest tree in urban settings (bearing in mind that it grows to significant size). With an eye to the future, golden larch should also be planted in large landscapes where it can reach its full stature. Its golden fall color is especially striking before a background of deep green conifers.

PLATE 110. 'Confetti' is a spectacular pink-and-white variegated selection of glossy abelia.

PLATE 111. Japanese momi fir is a statuesque tree that will thrive under conditions that no other true fir can tolerate.

PLATE 112. The glossy evergreen foliage of Japanese momi fir remains attractive year-round.

PLATE 113. Our native whitebark maple has brilliant fall color.

PLATE 114. The warm tan bark of *Acer triflorum* exfoliates attractively.

PLATE 115. Three-flowered maple's scarlet fall color can rival that of any other maple. (KIM TRIPP)

PLATE 116. *Aronia arbutifolia* 'Brilliantissima' is an exceptional cultivar of red chokeberry with vivid scarlet fruit.

PLATE 117. The lovely floral display of 'Brilliantissima' is as appealing as its ornamental fruit.

PLATE 118. The fall display of *Baccharis halimifolia* is created by silky hairs attached to the developing fruits.

PLATE 119. Both the white-fruited forms of *Callicarpa americana* and *C. dichotoma* have showy fall fruit displays.

PLATE 120. The common *Callicarpa americana* has uncommonly showy fall displays of bright purple fruit.

PLATE 121. 'Sundance' is a cultivar of *Choisya ternata* with lemon-yellow foliage.

PLATE 122. Native persimmon trees develop distinctive "alligator hide" bark.

PLATE 123. The vigorous, variegated thorny elaeagnus will thrive under almost any conditions.

PLATE 124. *Eleutherococcus sieboldianus* 'Variegata' (formerly *Acanthopanax sieboldianus*) is a beautifully variegated shrub that makes a lovely addition to a perennial or mixed shrub border.

PLATE 125. In areas where ash decline is not a problem, *Fraxinus americana* 'Autumn Glory' offers some of the most stunning fall color of any large tree.

PLATE 126. *Fraxinus americana* 'Autumn Purple' is a wonderful choice for exceptional fall color, in regions where ash decline is not a problem.

PLATE 127. *Gelsemium rankinii* covers itself with canary-yellow flowers in fall as well as spring.

PLATE 128. *Ginkgo biloba* is famous for its status as a "living fossil" and for its golden fall color.

PLATE 129. The fall color of our native *Hamamelis virginiana* can be equally dramatic in wild or garden settings. (KIM TRIPP)

PLATE 130. The unusual flowers of fall-blooming witch hazels emerge at the same time that fall foliage is reaching peak color, and so they may be overlooked at first. (KIM TRIPP)

PLATE 131. Flowers of fall-blooming witch hazels may persist past leaf-fall in some years, and then their unusual shape and appealing warm yellow color can be fully appreciated.

PLATE 132. The cherry-colored persistent sepals of the flowers of seven-son flower are even more showy than the creamy white petals.

PLATE 133. 'Mary Nell' holly has exceptionally glossy, finely toothed foliage.

PLATE 134. The mitten-shaped foliage of tulip poplar turns bright yellow-gold in autumn. (KIM TRIPP)

PLATE 135. *Nyssa sylvatica* 'Dirr's selection' is an unnamed selection by Dr. Michael Dirr made for its brilliant fall color.

PLATE 136. Black gum is an especially handsome, large native tree with dramatic red fall color. (KIM TRIPP)

PLATE 137. *Osmanthus fragrans* var. *aurantiacus* is the orange-flowered form of the fragrant tea olive; its flowers smell as sweet as those of the species.

PLATE 138. The deep red to burgundy fall color of sourwood (*Oxydendrum arboreum*) often sets off its late inflorescences.

PLATE 139. The mottled bark of Persian parrotia is a fascinating garden element.

PLATE 140. Persian parrotia develops excellent fall color in shades of red, orange, and yellow.

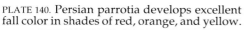

PLATE 141. 'Village Shade' is an exceptional selection of oriental photinia from the NCSU Arboretum, with attractive fruit.

PLATE 142. Photinia 'Village Shade' bears creamy white flowers in summer.

PLATE 143. Chinese pistache tree develops brilliant fall color even in hot climates.

PLATE 144. The autumn bloom of *Prunus subhirtella* 'Autumnalis' creates a surprising and delightful combination with the fall color of other trees. (ROBERT HYLAND)

PLATE 145. The light pink flowers of *Prunus subhirtella* 'Autumnalis' contrast beautifully with its own foliage. (ROBERT HYLAND)

PLATE 146. The foliage of the deciduous conifer, *Pseudolarix amabilis*, turns copper-gold before it falls.

PLATE 147. The fall color of a large golden larch is a magnificent, bold element in the autumn landscape. (KIM TRIPP)

PLATE 148. 'Apache' firethorn bears bright red fruit.

PLATE 149. Scarlet oak, *Quercus coccinea*, lives up to its name each fall.

PLATE 150. *Rhamnella franguloides* is a tough but little used small tree that develops lovely yellow fall color.

PLATE 151. The fruit of *Rhamnella franguloides* is ornamental.

PLATE 152. *Rhus copallina* turns a spectacular scarlet in fall.

PLATE 153. The fruit or "hips" of *Rosa villosa* are especially large, turning bright red-orange as they mature and persisting into winter.

PLATE 154. *Sapindus drumondii* is a dependable tree that develops soft but showy fall color.

PLATE 155. Korean mountain ash is a beautiful, disease-resistant tree with pink-red fruit displayed against yellow fall color that persists on the branches into the winter.

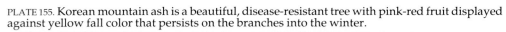

PLATE 156. 'Burnished Gold' is a striking cultivar of Japanese ternstroemia with bronze to copper-gold new foliage.

PLATE 157. 'Summer Snowflake' viburnum blooms prolifically from summer through fall.

PLATE 158. Tea viburnum not only has showy red fruits in fall but bears attractive flowers as well.

PLATE 159. The fruits of *Viburnum setigerum* color well in early fall.

PLATE 160. Tea viburnum fruits color early and hold their color beautifully well into winter.

PLATE 161. The fruits of Chinese date are ornamental as they ripen among the shiny green foliage and are quite delicious.

Pyracantha spp. / Firethorn, Pyracantha   PLATE 148

Clusters of brilliant orange-red fruit hang like tiny neon cherries along shrubby branches. Fine glossy, dark green foliage sets off their color to even greater intensity amid warm brown stems. But if you reach out to touch these bright fruits, you may get surprised by the sharp point of a nasty thorn. What devilishly attractive garden character is this? Undoubtedly a variety of *Pyracantha*, firethorn.

This very popular semievergreen shrub offers several species and dozens of cultivars. Exact size, hardiness, and fruit color vary among them all, but their basic characteristics are similar. Pyracantha can reach anywhere from 6–20 feet (1.8–6.2 m) in height, with a spread of about 10 feet (3.1 m). It is an open, somewhat unruly plant that grows very vigorously. Its stiff, upright branches are generally thorny, depending on species and cultivar. The foliage may be evergreen to semi-evergreen, depending on the variety and the region where it is grown.

Lacy white flower clusters shine around the branches from late spring through early summer. Firethorn flowers look and smell almost identical to hawthorn (*Crataegus* spp.) flowers, but they are smaller. Like the flowers of hawthorn, their fragrance is somewhat controversial. It is quite a heavy aroma and considered to be unappealing by many, although it is not as intense as the scent of hawthorn blooms. The flowers mature into beacons of fruit in the fall. The fabulous color of the fruit clusters adds a bold, vibrant note to the landscape that can be used in many ways. Unfortunately, birds love the fruit of many *Pyracantha* and will steadily eat them until they are all gone. However, some cultivars have fruit with much less bird appeal than others, so that the bold color of their fruit can be enjoyed through the winter.

Pyracantha is a tough plant and very drought resistant. It can get somewhat out of control without regular pruning (which can be done almost any time of the year except during flowering). It prefers dry sites in full sun but will tolerate partial shade, although flowering and fruiting will be reduced. Firethorn is a close relative of hawthorns and roses and is susceptible to many of the same diseases, including fireblight and scab, from which these and other members of the Rosaceae botanical family suffer. It is best to purchase cultivars with improved disease resistance and to use container-grown plants to avoid problems with root growth following transplanting. *Pyracantha* is readily propagated from either seed or cuttings. Seed is layered in a moist medium and stored at 40°F (4°C) for three months before sowing. Summer softwood cuttings are easily rooted under mist after treatment with rooting promoters. Cultivars are propagated vegetatively, by cuttings, to preserve their characteristics, while species and botanical varieties can be propagated by seed as well, as long as they are isolated to prevent hybridization with other *Pyracantha*. In other words, the pollen from one species can fertilize the flowers of another species to create "hybrid" seed that will grow into plants having the characteristics of both parents, and which are no longer truly one species or another.

Pyracantha coccinea, scarlet firethorn, is native to Italy and surrounding regions. It can be distinguished from other pyracanthas by its slightly serrate leaves. This firethorn is reliably hardy to zone 5 or 6, depending on cultivar. Some good cultivars of scarlet firethorn include 'Aurea', with yellow fruits; 'Lalandei', perhaps the most well known, an exceptionally hardy cultivar with orange-red fruit; and 'Thornless', with red fruit and no thorns.

'Cherri Berri', a cultivar of *Pyracantha fortuneana*, has exceptionally large, cherry-red fruit. Birds seem to be less attracted to this cultivar than some of the others. It has not been tested for hardiness extensively, but is likely hardy to zone 6.

Pyracantha koidzumii, Formosa firethorn, is a native of Formosa (Taiwan) with smooth-margined leaves. Significantly less hardy than *P. coccinea* or *P. fortuneana* (to zone 8), it will suffer damage at 0°F (−16°C). It has very dark green, glossy foliage that is especially handsome. Some good cultivars include the red-fruited 'San Jose', which is resistant to scab; 'Santa Cruz', which is also red-fruited and scab resistant and which has a unique prostrate habit; and 'Victory', which boasts extra-vigorous growth, arching branches, and excellent retention of its red fruit.

Some of the most desirable *Pyracantha* are hybrids developed by various plant breeding programs at universities, arboreta, and nurseries. The late Dr. Donald Egolf of the United States National Arboretum was renowned for his plant-breeding efforts with many different ornamentals, including *Pyracantha*. 'Mohave' is one of his first releases. It offers dense flowering, a heavy fruit load, good disease resistance, and low bird appeal, but is not reliably hardy below 0°F (−16°C). 'Apache' (Plate 148), one of his best, has slow, compact growth; exceptionally large, rich red fruit that are not a favorite of birds; and good resistance to scab and fireblight. 'Pueblo', another exceptional selection from this program, has vigorous growth, prolific orange-red fruit, and good disease resistance. Other commendable U.S. National Arboretum hybrid cultivars include 'Teton', with very upright habit and yellow-orange fruit; 'Shawnee', with good disease resistance and yellow fruit; and the low-growing 'Navaho', with orange-red fruit and excellent disease resistance.

Other good hybrid cultivars include 'Rutgers', released by Dr. Elwin Orton at Rutgers University; 'Red Elf', from Monrovia Nursery; and 'Gold Rush', from Washington Park Arboretum at the University of Washington in Washington state. 'Rutgers' is an exceptionally hardy selection (zone 5) with spreading habit. It offers orange-red fruit and has good disease resistance. 'Red Elf' is a dwarf with bright red, persistent fruit, and some disease resistance. It is hardy only through zone 7. 'Gold Rush' has prolific yellow fruit and shows good disease resistance.

Form, size, hardiness, and fruit color are all factors to consider when choosing *Pyracantha* for the landscape. Firethorn is not the best choice for very small or intensively planted areas because of its robust nature; however, given adequate space, *Pyracantha* can make a spectacular display, either massed or as a single specimen. It is wonderful when espaliered or trained up fences and other posts which then become covered in extravagant fruit each fall. Whatever cultivar you select, the vivid fruit of firethorn will add a burst of color to the fall garden.

Quercus coccinea / Scarlet oak   PLATE 149

Fall is the season when the scarlet oak, *Quercus coccinea*, really shines. The lobed leaves turn bright scarlet, creating large swaths of brilliant color that appear even more intensely red against the deep blue of autumn's sky.

Scarlet oak, like most oaks, is a large tree. It reaches 70–80 feet (21.7–24.8 m) in height and spreads to 50 feet (15.2 m) in width. When young, the tree's form resembles the pyramidal shape of its close relative, pin oak (*Quercus palustris*), which has similarly shaped foliage. As the scarlet oak tree develops, however, the canopy becomes rounded and more open. In the summer, the classic oak-outline leaves are a deep, glossy green. The leaves are very much like those of the more common red oak, *Quercus rubra*, with deeply incised and sharp-pointed lobes. Scarlet oak's acorns are a good source of food for wildlife. The acorns are about 1 inch (2.5 cm) long. Each acorn is almost completely enclosed by its furrowed cup.

Scarlet oak is native to the eastern half of the continental United States, where it can be found growing on light, sandy soils. It is hardy to zone 4. In the landscape, *Quercus coccinea* prefers a well-drained site. It is not as susceptible to foliar chlorosis as are many of the other oak species, and it has no significant pest or disease problems. Scarlet oak is not very tolerant of stressful sites and needs moderate, regular moisture. It can be difficult to transplant, so move balled-and-burlapped or container-grown trees, and make sure to stake trees securely and water regularly the first three seasons. Once successfully transplanted, scarlet oak is not difficult to grow and is well worth the initial attention.

Scarlet oak can be easily confused with the pin oak, which has very similar leaves. This is something to take note of when looking for the scarlet oak at nurseries, especially because pin oak is more readily available. *Quercus coccinea* is generally propagated from seed, which requires a cold pretreatment for a month. The cultivar 'Splendens' has especially glossy foliage but unfortunately is rarely available.

On your next constitutional on an autumn afternoon, when you feel the crunch and roll of acorns beneath your shoes, look up to see which oak has dropped its harvest for the squirrels and deer to collect. If you are lucky, it will be scarlet oak, spreading its limbs rustling with cardinal color against the clear, cooling sky. There's a way to bring this color to your garden, and that is to plant scarlet oak in a sunny spot with plenty of room to grow. It will reward you for years by bringing nature's autumn palette into your garden.

Rhamnella franguloides / Rhamnella   PLATES 150, 151

Rhamnella's vivid gold fall color is vibrant from acres away, and it positively glows in the garden, especially when sited next to other trees with bright red fall color. This adaptive small tree with year-round interest and good landscape character is another one of the unfamiliar Asian trees that deserve more attention and use in American gardens.

Common names are often nonexistent for unusual plants that have not yet reached the horticultural industry in any numbers. *Rhamnella franguloides* is such a tree, but don't let the awkward botanical name deter you from pursuing this unique and excellent plant.

Rhamnella is a small to medium-sized deciduous tree. It can ultimately reach 30 feet (9 m) in height with age in its native Korean environment, but will likely remain smaller in the lifetime of most American landscapes. It develops a rounded, well-branched canopy with lovely foliage of long, fine-toothed leaves, 6–8 inches (15.2–20.3 cm) long and half as wide. The leaves remain an unblemished rich, glossy green through hot summers and turn a show-stopping clear gold in the fall.

The warm gray bark is an asset in winter. Spring flowers are not especially showy, but they develop into oblong fruit with a brilliant range of colors that changes from yellowish to orange to brilliant red to black as the it ripens (Plate 151). The fruits ripen in midsummer and are dearly beloved by birds.

Rhamnella franguloides is likely hardy in the United States to at least zone 6. It performs best in full sun and thrives in a range of soils from heavy clays to well-drained loams. Full sun gives best shape, fruiting, and fall color, but the tree will probably also perform well in light shade. Rhamnella is propagated from seed or root cuttings. Seed is removed from the fruit pulp and given cold stratification. Root cuttings are taken in midwinter.

Adaptable small trees are stars in our demanding urban landscapes. There never seem to be enough selections to fill this niche, in courtyards, under wires, as screens, and as specimens for small properties. Rhamnella's tough, no-nonsense performance, handsome foliage, wonderful fruit, and handy size are an unbeatable combination. At the North Carolina State University Arboretum, *Rhamnella franguloides* is a pure gold player in the mixed border's west end. This is a perfect small tree for the mixed border or at the edge of a patio, and a fine candidate for the suburban yard or a corner of a courtyard. It may make a good small suburban street tree (could this be the next Bradford pear?).

Rhus spp. / Sumacs   PLATE 152

Sumacs' formal habit and large, divided leaves are attractive in summer, but it is in fall that they truly shine in the garden. Most sumacs develop intense fall colors (even as very young, small plants), in brilliant scarlets and orange. They are absolute showstoppers in any setting.

A number of sumacs, both native and exotic, offer wonderful garden characteristics for difficult landscapes. All sumacs for the garden are members of the genus *Rhus*, not to be confused with poison sumac (*Toxicodendron vernix*), a member of the same botanical family (but not the same genus) as garden sumacs, and a close relative of poison ivy (*Toxicodendron radicans*). *Rhus* species do not produce the toxic oils that can cause the allergic reactions many experience after touching poison sumac or poison ivy. Garden sumacs are friendly plants that will grow and thrive in conditions too stressful for other plants. They are worthy of greater use in many landscapes.

Sumacs are, in general, bushy shrubs of varying heights with divided foliage. The foliage are glossy green throughout the summer and give a brilliant fall color display of reds, wines, and golds. Hundreds of tiny green-white flowers are borne in soft, sometimes pendulous panicles, which ripen to masses of miniature scarlet, furry berries in the fall. These fruits often persist throughout the winter, adding color and dimension to the winter garden. Sumacs bear female and male flowers on different plants; female plants must be within range of males for good fruit displays.

All of the garden sumacs are hardy to zone 3 or 4. Most will grow well in dry, infertile, stressful sites as well as more favorable garden conditions. They will generally also tolerate heavy clay soils. Sumacs tend to sucker, creating large masses that can be effective for covering banks, cuts, areas stripped of topsoil, and other tough sites. Species are propagated from seed, following an acid treatment to allow water absorption and penetration of the seed coat by the young embryo plant and, usually, a cold treatment of several weeks. Cultivars are vegetatively propagated by root cuttings.

Perhaps the most well known sumac is *Rhus typhina*, staghorn sumac, a native of the eastern United States and into Canada around the Great Lakes and into Nova Scotia and New Brunswick. Named for its antler-shaped branches, staghorn sumac can be seen growing alongside highways, in old fields, and in other abandoned areas. *Rhus typhina* is a tall, rangy sumac reaching 20–30 feet (6.2–9 m) in height (or more) with equal spread. The branches are covered with a light brown pubescence, reminiscent of the velvet on the antlers of stags. The foliage has spectacular fall color. Perhaps not as tolerant of poorly drained soils as some of the other sumacs, it is nonetheless adapted to a wide range of soil conditions. 'Laciniata' and 'Dissecta' are especially beautiful cultivars of this plant, and very difficult to tell apart. Both have heavily divided leaflets that give the foliage a fernlike, incredibly graceful appearance. 'Dissecta' is supposedly even more deeply divided than 'Laciniata', but few people can see the difference. Both of these cultivars are female plants with excellent red and orange fall color. 'Laciniata' is probably the more readily available of the two.

Fragrant sumac, _Rhus aromatica_, can be found growing in dry, rocky, open areas in much of the eastern and central United States, and in the Great Lakes area of Canada. Named for the pungent odor of its crushed leaves, fragrant sumac is a low shrub, generally smaller than staghorn sumac, reaching a maximum height of 5–6 feet (1.5–1.8 m) with equal or greater spread. The summer foliage of this sumac is very dark green and glossy, more handsome than most sumacs. The fall color, however, is not as dramatic as others in hot climates. Fragrant sumac has the added charm of bright yellow flowers in the spring. Cultivars include 'Grow-Low', a very low, spreading plant, and 'Green Globe', a very dense, rounded shrub.

Another North American native is _Rhus glabra_, smooth sumac. Very similar to staghorn sumac, it can be distinguished from its cousin by its smooth, hairless stems. Smooth sumac is also somewhat shorter, reaching 10–15 feet (3.1–4.5 m) in height, and may have more finely textured foliage than staghorn sumac. Fall color is excellent. The cultivar _Rhus glabra_ 'Laciniata' is nearly identical to _Rhus typhina_ 'Laciniata', except for its smooth stems.

Yet another native, _Rhus copallina_ (Plate 152), shining or flameleaf sumac, is a large and vigorous sumac, especially in the Southeast, where it is quite rampant and can easily grow out of control. Shining sumac can easily reach 30 feet (9 m) in height. It becomes spreading and irregular as it ages, and the branching can be quite interesting on an older plant. It forms large, spreading colonies and is difficult (if not impossible) to use on small sites, but it is a beautiful plant with spectacular carmine fall color.

Of all the sumacs, Chinese sumac, _Rhus chinensis_, is perhaps the best behaved as a garden plant. Native to China and Japan, this species is not quite as voracious a spreader as other sumacs. It forms a large, rounded mass, reaching 25 feet (7.6 m) in height with a spread of 15 feet (4.5 m), and can be pruned to a multistemmed tree form. The unique character of Chinese sumac is its late bloom time. Unlike the other sumacs, which bloom in spring, Chinese sumac flowers profusely in late summer or early fall. Thousands of almost invisibly small, whitish flowers are borne in large, dramatic panicles. These flower clusters make a beautiful display as they droop gently against the plant's soft green foliage. 'September Beauty' is a fabulous cultivar with remarkably large flower clusters, which can be seen at the North Carolina State University Arboretum along the southern border of the west arboretum just beyond the Japanese garden.

Garden sumacs offer graceful summer beauty, intense fall color, and winter interest for even the most difficult of landscapes. Sumacs are excellent in informal mixed borders, for bringing bright fall color right into the border itself. The large species can be good substitutes for small trees in very tough conditions where other small trees can't survive. They are great tall groundcovers on dry banks and in other challenging sites, but can also be well-behaved in small gardens if the less rampant species are used. This group of shrubs can add character to many more American gardens than they do now—an oversight that can easily be remedied by including many of the sumacs throughout your own garden.

Rosa villosa / Apple rose   PLATE 153

A special standout in the world of roses is _Rosa villosa_, apple rose, an unusual species that brightens the autumn garden with extraordinarily large—2–4 inches (5–10.1 cm) long—pear-shaped hips. The hips develop into brilliant scarlet-orange fruits that dangle from the branches like flashy jewelry. They glow with color from across the garden and are an absolute magnet for children and curious visitors.

Rosa villosa is a rose of central European origin that is sometimes listed as *Rosa pomifera* or *Rosa mollis*. It is a medium-sized, deciduous shrub with attractive, somewhat glossy green foliage of the basic rose variety. Apple rose can develop lovely soft yellow fall color (if it has not defoliated from leaf spot or other rose foliar diseases). It is moderately thorny, multistemmed, and will reach in the neighborhood of 4–6 feet (1.2–1.8 m) in height with some age.

Apple rose has single, blush-pink blooms in late spring and early summer. The flowers mature in late summer and early fall into what is probably the most striking fruit of any rose. The bright fruit persists well into winter.

Rosa villosa prefers moist, well-drained soil but will tolerate moderately heavy clay and moderate drainage. It is hardy to zone 5. Although subject to some of the rose foliar diseases, apple rose will still produce a good crop of vivid hips year after year, in spite of disease injury. *Rosa villosa* is propagated from softwood cuttings rooted under mist in summer.

Cultivars of apple rose selected for various fruit qualities are available in Europe, but they are rarely available in the United States. The species *Rosa villosa* can be purchased through specialty mail-order nurseries, especially those dealing in "old roses" and alternative rose selections.

The unbelievable fruit of the apple rose will liven up any border. The plant makes a good corner for an evergreen hedge, and it also takes to life in a large patio container. A rose by any other name would still smell as sweet, but only *Rosa villosa* will give gardeners the treat of brilliant orange-red fruit shimmering through Indian summer.

Sapindus drumondii / Western soapberry   PLATE 154

One of the most challenging aspects of landscaping our cities is planting to combine seasonal interest with survivability. Although some of the best plants and shrubs succumb to the stresses of urban blight, enough stalwart yet still attractive plants can be found to fill all four seasons of the city with flowers and fragrance, bright fruit and brilliant foliage. One tree that contributes beautifully to the urban landscape is *Sapindus drumondii*, western soapberry. In the fall, this tough tree literally shines, its divided foliage turning a blazing clear yellow. The intense color—never garish in spite of its brightness—and fine texture are an outstanding combination.

Western soapberry is a North American native that can be found growing throughout the southern half of the Midwest, from Missouri to Kansas, south through Texas and into Mexico and east into Louisiana. It is a mid-sized, deciduous tree, reaching 30–40 feet (9–12.4 m) in height with an almost equal spread to its neat, rounded crown. The foliage is soft and divided, giving it a fine texture for such a large tree. Leaves emerge a light green in spring and darken to a glossy grass-green in summer. The foliage remains blemish-free and quite handsome throughout the growing season, and the yellow fall color is glorious. In late spring, creamy flowers in loosely branched, upright clusters, 8" long, wave atop the branches. The flowers mature into rusty yellow soft fruits. The fall color is especially eye-catching against the gray bark, which sometimes sheds in plates, revealing warmer tones of brown in the newly exposed underbark.

Western soapberry is extremely tolerant of urban sites and is easily transplanted. It will grow in full sun or a little shade, and thrives in a range of soils, from heavy clay to very dry. It is extremely drought tolerant and will take the abuse of transplanting into a tough site with minimal watering. The tree has no serious disease or pest problems and is hardy to zone 5. *Sapindus drumondii* is usually propagated from seed. The leathery, waxy fruit covering is removed, and the seed is pretreated with moist storage for two months in cold temperatures to induce germination.

Other species of *Sapindus* are also in cultivation. One is the nonhardy *S. saponaria*, whose fruits are crushed for use as a soap substitute in its native tropics because of the lather they produce. The Latin root of the name *Sapindus* is a combination of *sap* for "soap," and *indus*, which indicates use by the "Indian" or indigenous people of the West Indies, where the fruits are used in this manner.

Western soapberry is a potential hero for the modern urban landscape. Sadly, it is generally unavailable in the eastern half of the United States and will require hunting to find. While it is a bit large for patio or townhouse gardens, it makes an excellent street tree or urban park specimen. It could be far more widely used in many of the new urban spaces, such as parking lots, walkways of shopping centers, and urban pedestrian malls, as well as new residential areas. Its grass-green leaves, creamy flowers, and golden sunburst of fall color remind us all that the seasons still turn, even right outside the high-rise.

Sorbus alnifolia / Korean mountain ash   PLATE 155

The foliage of Korean mountain ash is uniquely beautiful all through the year, but its most dramatic display is in autumn, when the leaves become bright orange. Bright berries in cherry-red to vermillion provide another part of the spectacular fall show of *Sorbus alnifolia*. The fruits are absolutely spectacular against the fall foliage and this tree's elegant, smooth gray bark.

Most *Sorbus* species are ravaged by root rots and show increased susceptibility to the many pests and diseases that already plague many of its rose family relatives; however, *Sorbus alnifolia*, a species from China, Japan, and Korea, stands far above the others in terms of trouble-free status. It also just happens to be one of the most striking trees in the fall landscape.

Korean mountain ash is a medium to large tree reaching 40 feet (12. 4 m) in height with a 20-foot (6.2-m) spread. The crown becomes generously rounded as the tree matures from its oval habit in youth. Korean mountain ash has single, entire leaves, reminiscent of beech leaves (*Fagus* spp.); many other *Sorbus* species have divided leaves similar to ash (*Fraxinus* spp.) or pecan (*Carya illinoiensis*). The foliage of *Sorbus alnifolia* is beautiful all through the year, emerging a velvety emerald which darkens to a rich grass-green and finally colors dramatically in autumn. Korean mountain ash flowers in late spring, bearing flattened clusters of delicate white flowers. These flowers mature into the striking show of colored fruit, which the birds also seem to enjoy.

Sorbus alnifolia shows very little damage from fireblight, borers, scab, and the host of other diseases possible with other *Sorbus*. It is very cold-hardy and can be grown safely to zone 4. Korean mountain ash is well adapted to diverse soils, including heavy clay and loamy sands. It does not tolerate high levels of air pollution well, and so makes a poor choice for inner-city spaces. The tree is propagated from seed, which requires three months of moist, cold pretreatment before it will germinate.

The *Sorbus alnifolia* cultivar 'Redbird' offers remarkably persistent, rose-red fruit; 'Skyline' is broadly upright and columnar. Numerous cultivars of other *Sorbus* species are plentiful and should not be confused with Korean mountain ash.

Korean mountain ash makes an unusual and beautiful lawn specimen tree and is perfect for suburban neighborhoods where the overly quiet yards cry out for showy relief from miles and miles of foundation plantings. While *Sorbus alnifolia* is sensitive to air pollution, it is otherwise a relatively tough plant. The rewards of Korean mountain ash in fall color and fruit far outweigh the efforts to find this tree and site it away from the heart of the city.

Ternstroemia gymnanthera / Japanese ternstroemia   PLATE 156

Most conifers and many broadleaved evergreens need full sun for most of the day to perform well. In southerly areas of the United States, that rule can often be bent because of the increased intensity of the light relative to northern areas. Many evergreens that would need full sun all day further north do just fine, thank you, with only half a day of sun—and in fact are often improved with some relief from the afternoon blast. But that is not the same issue as the challenge of full shade or heavy shade for most of the day. The selection of evergreen woody plants is quite limited under these conditions. For this reason, a handsome evergreen that prefers significant shade—a plant such as *Ternstroemia gymnanthera*—is a real treasure in the garden. Add the bonus of a crop of red cherrylike fruits in fall, and this broadleaved shade-lover is a winner.

Japanese ternstroemia is a mouthful of a name for a plant that is unassuming and easy to use in the shaded landscape. Native to Japan, Korea, and China, this broadleaved shrub is a botanical relative of camellias, although it does not bear an overwhelming resemblance to the familiar ornamental camellias. This is a large, well-branched, upright shrub, taller than broad, to 8–15 feet (2.4–4.5 m) in height with half the spread. The foliage has a rounded, oval shape, with glossy, leathery appearance and texture. The burnished leaves are olive-green, a deep, quiet color that is difficult to find in most woody plants. The leaves are arranged in clusters that add an interesting, almost layered look to the foliage. The color of the foliage varies from the red-bronze of new spring growth to rich, glossy olive in summer, and wine and moss-green in winter.

Japanese ternstroemia produces subtle clusters of small, creamy white flowers in late spring. The flowers mature into ornamental fruit in early fall. Japanese ternstroemia's fruit superficially resembles cherries in shape, size, and color. The fruits ripen to a deep red and dangle from long pedicels but are less fleshy and more oval than cherries. The visual effect is delightful.

Japanese ternstroemia is hardy to zone 7 and will suffer foliar damage and twig kill in severe winters. It does best in shady gardens but also tolerates full sun. The shrub will thrive in a range of soils from light sands to all but the heaviest clays. *Ternstroemia gymnanthera* is relatively slow-growing, but not excessively so. It responds very well to pruning and shearing and so can be maintained as a low hedge or screening plant in the shade.

Ternstroemia is generally propagated from seed collected in the fall and planted immediately; the seed germinates the following spring. Because plants are grown from seed, there is often great variability among individuals in foliage color, uniformity of habit, and other characteristics. Propagation is also successful from cuttings, which are taken in the early fall and rooted under mist.

There are very few cultivars. 'Variegata' has leaves marbled with white and silver, and it develops tinges of pink with cold weather. 'Burnished Gold' (Plate 156) has bronze-gold new foliage overlaid with burgundy, giving the leaves a unique appearance of aged velvet. Tom Dodd Nursery in Alabama offers two other interesting and as-yet-unnamed forms—a purple-foliaged clone and a handsome dwarf form. *Ternstroemia gymnanthera* is often confused with the virtually identical *Cleyera japonica*, Japanese cleyera, to the point that almost all plants sold as *Cleyera* are actually *Ternstroemia*.

Japanese ternstroemia can be the answer to the prayers of many gardeners—the solution to your shade garden's challenge. It is a beautiful specimen in the corner of a shade garden where it adds Japanesque character, especially against a bamboo structure. It makes a fine hedge for shade and will accept almost any shape of pruning. In the fall garden, the dangling fruits are a special treat. Plant *Ternstroemia* near the walk that ambles through the shadiest part of your garden, so that its wonderful foliage and fruit can be easily seen.

Viburnum plicatum var. *tomentosum* 'Summer Snowflake' / 'Summer Snowflake' viburnum

PLATE 157

Few flowering shrubs are showy during the transitional period from summer until the show of fall foliage, and those select woody plants that do flower at this season are thus extremely valuable, not to mention quite delightful, in the garden. *Viburnum plicatum* var. *tomentosum* 'Summer Snowflake', a selection of doublefile viburnum, is one of those welcome shrubs still in bloom in early fall. Released by the University of British Columbia Botanical Garden in Canada, this viburnum combines the graceful bloom character of other doublefile viburnums with an extended flowering period. In the warmer parts of the United States, this lovely plant blooms continuously from May into November!

Doublefile viburnums in general are deciduous, woody shrubs reaching 7–10 feet (2.1–3.1 m) in height, with a somewhat wider spread. The shrubs have a rounded habit. Large, deep green leaves emerge in spring, each with a distinctly toothed edge and significant creases along the veins, adding textural interest. The foliage is handsome in its own right, but when combined with the flowers, doublefile viburnum is one of the most beautiful plants in the landscape. The showy part of the flower clusters of doublefile viburnum are pure white. The unique, flat-topped flower clusters are made up of two kinds of flowers: an outer ring of showy sterile blooms, each with four petals (in the shape of *Hydrangea* blossoms), that surrounds a central disk of tiny fertile flowers that look like minuscule, creamy green beads. The flower clusters sit on top of the branches like floating bridal caps. Eventually the flowers mature into small oval, carmine-colored fruit that ripens to black, but the fruit is usually eaten by birds before it can make a significant show.

There are numerous cultivars of doublefile viburnum, almost all of which are excellent plants that bloom in April through June. 'Summer Snowflake' is one of only two doublefile viburnum cultivars that continue to flower through autumn.

In early fall, when flowering is still profuse, the foliage of 'Summer Snowflake' begins turning a rich mahogany-wine color. The combination of the white flower clusters scattered among this turning foliage is magnificent, like lace antimacassars on antique velvet, and brings wonderful character to the garden.

Doublefile viburnum in general requires moist, well-drained soils in full sun or partial shade, with full sun giving the best flowering and fruiting. It is hardy to zone 5 and is a relatively trouble-free plant, but needs some irrigation in areas with summers where it does not rain. 'Summer Snowflake' flowers dependably through the spring, summer, and fall, with good fall color but little fruit production. It performs well in both heavy clay soils and well-drained loams and needs full sun for best flowering. All doublefile viburnums, including 'Summer Snowflake', can be easily propagated from stem cuttings taken throughout the growing season and rooted under mist.

'Summer Snowflake' viburnum brings the elegance of burgundy velvet and white lace to the early fall garden. It is a stunning choice for a mixed shrub border or a low screen, a patio specimen, a mass planting in a lawn, or to soften hard edges and awkward corners in the garden.

Viburnum setigerum / Tea viburnum   PLATES 158–160

Many viburnums, such as the European snowball bush (*Viburnum opulus* 'Roseum') and doublefile viburnum (*Viburnum plicatum* var. *tomentosum*), have long been popular for their beautiful white flowers and reliable foliage, but there are lesser known viburnums that deserve more attention for their own wonderful character. *Viburnum setigerum* is just such a plant.

In the fall, the branches of tea viburnum arch gracefully toward the ground under the weight of their vermilion fruit (Plate 159, 160). The bountiful display is a feast for our eyes and a literal feast for birds.

This large deciduous shrub is native to China, where its leaves were used to make tea. It grows 8–12 feet (2.4–3.7 m) tall and 6–8 feet (1.8–2.4 m) wide, assuming a vase shape and losing its lower leaves as it matures. The foliage is a soft gray- to blue-green and remains attractive through the summer.

In mid-May, white saucer-shaped flower clusters dapple the foliage in an understated show for spring. In early fall, the oval, 0.75-inch (1.9-cm) fruits develop and ripen to a brilliant cardinal-red that persists through late fall. Some plants will develop purple-wine fall color, but this trait varies and is not the shrub's primary attraction in the landscape. The striking feature of tea viburnum is its autumn fruit display.

Tea viburnum is hardy to zone 5. It does well in moist but well-drained soils, and fruits most prolifically in full sun, although plants in shade may still provide a good fruit display. It has no serious disease or pest problems. Rooting of summer softwood cuttings is the preferred method of propagation, but the shrub can be grown from seed as well.

A unique cultivar of tea viburnum called 'Aurantiacum' has all the qualities of the species but produces orange fruit that makes an outstanding landscape statement in the fall. One caveat, however: You will probably have to fight for the fruit of 'Aurantiacum' with the birds; they seem to particularly love the orange-colored fruit.

While many viburnums give a showy fruit display in areas with cold winters, fruiting is often reduced in regions with hot summers and warm winters. This is not true of tea viburnum, making it an ideal choice for mild-winter regions.

Fruiting tea viburnum gives an incredible display when used in a shrub border or as a hedge. A grouped planting has the added advantage of camouflaging the legginess that develops as the shrubs mature.

Ziziphus jujuba / Jujube tree   PLATE 161

The deliciously different fruit of this unusual ornamental tree offers a unique fall display in the garden, provides food for wildlife, and serves as a treat for people as well. Also known as Chinese dates, the oval, 1-inch-long (2.5 cm) fruit of *Ziziphus jujuba* is somewhat reminiscent of true dates (the fruit of the date palm, *Phoenix dactylifera*). The fruit is not just edible, but sweet and tasty. Jujube fruit is somewhat drier-textured than most dates but just as flavorful.

Jujube tree is a handsome plant of rounded habit, with very glossy, rather fine-textured foliage. Sometimes seen as a multistemmed large shrub, the tree can eventually reach 25 feet (7.6 m) in height, but it usually grows to 8–10 feet (2.4–3.1 m) in the landscape. Small yellow flowers borne in the spring are not especially showy but are delicately pretty. In fall, the leaves generally turn an appealing clear yellow, but this feature is somewhat variable from tree to tree.

Fruits are ripe in autumn when they have darkened to a deep maroon or brown-black color. Dr. J. C. Raulston reports that he eats them "steadily as they ripen." The fruits are a staple food in the Mideast. They can be dried and stored for later use.

Jujube is a buckthorn (*Rhamnus* spp.) relative native to southern and eastern Asia and parts of southern Europe. It is fully hardy to zone 6. The plant does best in well-drained soils but will also tolerate heavy clay. Jujube will flower and fruit most prolifically in full sun, but is extremely stress-tolerant and will also perform well even in severe heat and drought. The species is propagated by seed, which requires a pretreatment of three months of warmth followed by three months of cold temperatures for good germination. Cultivars are propagated by grafting onto seedling rootstock.

Seedling trees are available, but the size and quality of their fruit is not as reliable as that of cultivars. The fruit of cultivars can be twice the size of the fruit of seedling trees, in the range of 2–4 inches (5.1–10.2 cm) long. Both 'Lang' and 'Li' offer larger fruit than seedling trees.

Jujube is a playful specimen tree, well suited to gardens with limited space. Its tolerance of heat and drought also makes it a good choice for tough urban landscapes.

Winter

Acer palmatum 'Sango-kaku' / Coral-bark Japanese maple   PLATES 162, 163

The beautiful bark of woody plants offers a surprising palette of hues and textures for the winter garden. Especially breathtaking in winter is the silken coral-colored bark of *Acer palmatum* 'Sango-kaku'. The growth habit of the tree is more upright than that of many other Japanese maple selections with narrow branch angles and ascending branches sweeping upward to create a brush of color. 'Sango-kaku' is a dramatic feature in the winter landscape, with its glowing pink limbs stretching upward against the dark greens and grays of other trees.

A cultivar of the familiar Japanese maple, *Acer palmatum*, 'Sango-kaku' was selected in Japan primarily for its brilliant, coral-red winter bark. During cold weather, the color intensifies to an almost fluorescent salmon. Winter is considered the peak season of interest, but the tree is appealing in all four seasons of the year. The small, refined, palm-shaped leaves of Japanese maple are veritable horticultural icons, and the leaves of 'Sango-kaku' are no exception, with their narrow, pointed lobes. Like many Japanese maples, this cultivar's leaves emerge tinged with bright red in the spring and then turn a uniform jade-green for the summer. In the fall, the foliage turns yellow-gold with burnished overtones, creating a wonderful contrast with the coral bark as it begins to intensify toward its winter depth of color.

Native to eastern Asia, Japanese maples in general are good, tough plants that bring great beauty to the landscape year-round. The many forms range from dwarf varieties 2–3 feet (0.6–0.9 m) tall to large specimens with magnificent architectural branching supporting a 30-foot (9 m) crown. Coral-bark Japanese maple will reach 20–25 feet (6.2–7.5 m) in height with a nearly equal spread. It is reliably cold-hardy to zone 6, and may survive in zone 5 in protected microclimates. Japanese maples, including coral-bark 'Sango-kaku', prefer moist, loamy soil but will thrive in a range of soil conditions, from sand to heavy clay, and are quite tolerant of dry periods. They have no significant disease or pest problems.

Japanese maples perform well in full sun to relatively dense shade; color develops best in full sun and habit is more dense and compact than in shade, where the trees tend to be greener and a bit more open and "stretched." 'Sango-kaku' will generally have the boldest bark color if grown in the sun. Color is best on the youngest, most vigorous growth, but even older plants develop striking color in winter. Like most named forms of Japanese maple, 'Sango-kaku' is propagated by grafting young branches of the cultivar onto seedling rootstock in winter. Japanese maple can also be propagated by rooted cuttings, but it is a difficult and a slow process.

'Sango-kaku' is sometimes mistakenly called 'Corallinum' (and vice-versa), but they are very different plants. 'Corallinum' also has salmon-hued bark, but the color is not as intense as that of 'Sango-kaku'; this cultivar was named for its coral-colored spring foliage. 'Sango-kaku' is also occasionally listed as 'Senkaki', 'Coral Tower', or 'Cinnabar Wood Maple', but 'Sango-kaku' is the most common and correct name. The many Japanese names of all of the cultivars of *Acer palmatum* and related species can be a nightmare for the uninitiated to sort out. The standard reference, *Japanese Maples*, by the renowned expert J. D. Vertrees (Timber Press, 1987) is not only an excellent reference but an accessible, practical help.

Acer palmatum is the symbol of the North Carolina State University Arboretum because of its great beauty and durability, four-season interest, and well-adapted performance in a range of climates. Many different Japanese maples are planted throughout the Arboretum, including a number of 'Sango-kaku' trees. In the Arboretum's winter garden, and along the vine lattice, coral-bark Japanese maples create spectacular horticultural art in combination with all the different colors and forms of the winter landscape.

The bold color and vertical lines of coral-bark Japanese maples are an eye-catching note when used among evergreens, and they make bold combinations with plants that have gold or yellow twigs or foliage, such as many gold conifers or the yellow-twigged dogwood (*Cornus sericea* 'Flaviramea'). The flash of coral leaps to the eye, a welcome brilliance in the winter garden, and an incomparable framework for the flowers and foliage of other winter-interest plants.

Alnus spp. / Alders   PLATES 164, 165

After leaf-fall, when winter has truly settled in, our eyes become sensitized again to the subtle browns and grays of deciduous trees. Then the little woods and thickets of wild places become meditative works of art—interwoven trunks and branches subtly arrayed in blended earth tones, striped with sunlight and shadow. In the midst of such mental and visual quietude, the finely branched crowns and dangling catkins of alders are a delight. The winter display of alder catkins is not especially showy, but it adds valuable color to the landscape at that time of year. In late winter, yellow-brown or red-brown male flowers are borne on long, graceful catkins. Female flowers are borne on stubby, purple-brown, conelike "strobili," which persist through the following winter. Both male and female flowering structures are ornamental, adding charm and character to the tree's winter silhouette as they dangle from the limbs, dancing to the slightest breeze like silent wind chimes.

The genus *Alnus*, a close relative of the more familiar birch (*Betula*),with about 35 species native to various parts of the world, includes many exceptional garden and landscape plants. The alders range in size and habit from large trees to shrubby, multistemmed plants. Most have warm, light brown bark and ornamental branching patterns, providing special winter interest.

Alders are relatively tough plants, with a few subject to some insect or disease problems. They prefer moist sites and will tolerate flooding and submergence, but they will often thrive in dry soils as well. Growth is good in both full sun and part shade, but flowering and fruiting are best in full sun. Almost all of the alders have the attribute—unusual among woody plants—of being able to "fix" nitrogen with their roots, in the manner of legumes. This capability makes them excellent plants for poor soils and waste places. They are also among the most tolerant of woody species for wet and flooded soils, with species found growing along watercourses and wetlands in North America.

Alders can be propagated from fresh seed, sown directly, or from dried seed sown following a cold pretreatment of 3 months at 40°F (4°C). The germination rate is often low with alders. Vegetative propagation from cuttings can be reliable, if very soft wood is used. Some success has been reported with softwood cuttings taken in spring, treated with rooting promoter, and rooted under mist. Cultivars are generally propagated by grafting onto species rootstock.

In spite of their beauty and utility, alders are not commonly grown in most nurseries and gardens. They have a general reputation for being disease-prone and ratty in the landscape. While this is true for a few species—including some of our native species—it is not true for a number of exceptional species, especially the Asian alders. It seems that some of the shrubby North American alders have given the entire genus a bad name. This is a situation that needs to be changed, for alders have a great deal to offer modern gardens and landscapes. Among the many alders that you may encounter, several stand out as particularly gardenworthy.

Alnus cordata, Italian alder, is one of the least common and most attractive alders. Native to Italy, this alder has an eye-catching flame-shaped silhouette, and rich green, glossy foliage. It grows to be a large tree, reaching 50 feet (15.2 m), with upright branching. It is hardy to zone 5 but can be subject to canker and powdery mildew.

Alnus firma (Plate 165) and *A. japonica* are two of the species native to Japan. They have slender, upright habits and oval foliage that is more narrow and refined than that of the commonly grown European alder. Both make very graceful small trees that can reach 12–25 feet (3.7–7.5 m) in height with a relatively narrow spread of 8–12 feet (2.4–3.7 m). Both can grow as multistemmed or single-stemmed trees or large shrubs, depending on environment and cultural practices. Seedlings of both species usually develop into small trees. They are both reliably hardy through zone 7, with *A. japonica* hardy into zone 6 as well, but cold hardiness may vary with provenance of the seed parent plants, and neither remain attractive when grown in areas where they are at borderline hardiness.

Alnus glutinosa (Plate 164), common (or European) alder, is native to parts of Asia, Africa, and Europe, but has naturalized extensively along watercourses in North America. It can grow to be a large tree (Plate 164), to 80 feet (24.8 m), but is more often seen as a multistemmed small tree in cultivation and even as a shrub in the wild. The rounded leaves, as large as 4 inches (10.1 cm) across, are a shiny dark green and remain good-looking all summer. There is little fall color development. A few cultivars of *Alnus glutinosa* are available. 'Aurea' is a relatively slow-growing cultivar; its new foliage is gold. 'Charles Howlett' has variegated, non-uniformly shaped foliage. 'Imperialis' and 'Laciniata' have cut-leaved foliage. 'Pyramidalis' has a striking upright, columnar habit of growth. Hardy to zone 3, common alder is among the least tolerant of the alders to heat and humidity.

Alnus hirsuta, Manchurian alder, is native to Asia, and is perhaps the most beautiful alder. It is a large tree, to 60 feet (18.6 m), with a rounded oval crown and intricate branching hung with prolific catkins and strobili. The foliage is olive-green all summer and turns burnished yellow-tan in fall. Manchurian alder is hardy to zone 5. It is tolerant of hot summers and has no significant pest or disease problems.

Alnus incana, white alder, is a European species, very common in its native Europe but less so in the United States. It is a large tree, to 60 feet (18.6 m), with a broadly oval habit, and somewhat dull, medium green, tapered foliage. White alder is hardy to zone 3. There are a few cultivars of this species. 'Aurea' has yellow new foliage and bright yellow-orange male catkins. 'Laciniata' is a cut-leaf form with exceptionally deep lobing. 'Pendula' is a beautiful weeping form.

Alnus rugosa, speckled alder, is native throughout the northeastern United States and across most of Canada. It is hardy to zone 3. It is a shrubby, multistemmed plant or small tree to 20 feet (6.2 m) with dark, murky green foliage, more coarse in habit than the previously described alders. It is best used in naturalized situations.

Alnus serrulata, hazel alder, is also native through much of the eastern seaboard. Its habit and appearance are similar to *A. rugosa,* but its foliage is slightly different. It is most pleasing in naturalized settings, where the branches and catkins are a fine addition to the winter landscape. Hazel alder is hardy to zone 5.

All alders are good choices for poor, infertile soils and will tolerate wet soils. The shrubby species are useful for naturalized plantings along streams and ponds and in wet bottomlands. The tree-form species make some of the most effective shade and lawn trees available. The cut-leaved cultivars make good lawn trees, as their dappled shade is not as dense.

The elegantly branched crowns and delicate, warm-toned ornaments of the many gardenworthy alders are appealing in the midst of winter's mental and visual quietude. Manchurian alder is especially reliable and very appealing when the leaves have fallen, exposing the silver trellis of branches hung with last year's strobili—as if a sculpture were revealed to the winter light.

Buxus spp. / Boxwood   PLATES 166, 167

The structural elements of a garden create the garden's foundation—an underlying sense of formality or informality, ordered geometry or fascinating fluidity. We often think of the garden's structure as resting on hardscape features, such as stone walls, brick patios, and curving paths, but excellent structure can also be built by plants. Common boxwood, *Buxus sempervirens*, from northern Africa and southern Europe, and littleleaf boxwood, *Buxus microphylla*, from Japan and Korea, are two reliable foundation plants that give consistent, year-round structural dimension in the landscape. The fine-textured foliage and formal silhouette of boxwood can make an important structural element in the winter garden when deciduous plants have lost their leaves.

Gardeners on both sides of the Atlantic make ardent statements about the famous "English box" or "American box," when neither of the horticulturally important species is native to either England or America. "English boxwood" is *Buxus sempervirens* 'Suffructicosa', a slow-growing, mounding, fine-textured, centuries-old selection widely used in England, while "American boxwood" is *B. sempervirens* 'Arborescens', a somewhat more coarse-textured form that will eventually reach tree-type proportions.

Boxwood is one of the oldest known garden plants. Records of its use date back to 4000 B.C.E. in Egypt. Boxwood has been the traditional plant of formal English gardens and many French parterres. It has been very important in early American formal gardens, such as those restored in Williamsburg, Virginia, and it is preserved at a number of historically important Presidential homes and gardens. But boxwood is not only a historical plant. It offers a great deal to modern gardens, in both formal and informal roles.

Both *Buxus sempervirens* and *Buxus microphylla* are relatively slow-growing, evergreen shrubs with a billowy habit created by dense, fine-textured foliage. The dark, rich green leaves of boxwood are only 0.25 to 1 inch (0.6–2.5 cm) long and half as wide, but there are so many leaves packed so closely that the plants appear to have a nearly continuous surface from even a short distance away. One way to tell common boxwood from littleleaf boxwood is by the somewhat unpleasant odor of common boxwood foliage and stems when bruised. Even when lightly brushed, the aroma of common boxwood has been compared to that of a cattery.

Both species are hardy throughout zone 6 and warmer parts of zone 5, but the plants will show some winter burn in severe winters, especially if grown in a sunny site. There are many approaches to avoiding winterkill on boxwood, from simple burlap covers to elaborate frames covered with high-tech material. A loose wrap of burlap, tied to keep it from blowing and rubbing the foliage, can help, but be careful to remove the cover as the weather warms in spring and before bud-break. There is some variation in cold-hardiness among the numerous cultivars; exceptions are noted in the cultivar descriptions that follow. Boxwood generally prefers partial shade and cool roots in moist, well-drained soils; however, it can also do well in clay soils and full sun with plentiful water in times of drought.

Boxwood is a reasonably adaptable plant, but it is subject to some pest and disease problems. In general, if boxwood is sited well, pests and diseases are not serious and the plants remain healthy and vigorous for the extent of their long life. Occasionally, an individual plant may need replacing after a decade or so. Some potential problems include nematodes and root rot, which can gradually kill the root system. These root problems are usually indicated by orange- or rusty-colored foliage in one area of the plant (where the analogous roots are in trouble). Various insects attack the foliage. Regional extension agents can help with appropriate control measures if pests become serious problems.

Boxwood is readily propagated from cuttings taken any time of the year and rooted under mist. It can also be propagated from seed, but vegetatively propagated cultivars are superior to seedlings, which may vary in their character.

A huge number of cultivars of both species are available, offering a broad range of sizes and shapes, and including some variegated forms. Perhaps the most extensive collection of boxwood cultivars available in the country for public viewing can be found at the University of Virginia Blandy Experiment Station in Boyce, Virginia. There is also an extensive collection of cultivars at the National Arboretum in Washington, D.C.

There are many good cultivars of littleleaf boxwood. 'Helen Whiting', named after a well-known plantswoman and boxwood aficionado, is a plant with foliage of excellent quality. 'Kingsville Dwarf' is a very low growing form that can be sheared to make an interesting tall groundcover. 'Morris Midget' is a dwarf, formal, round cultivar with light green foliage and very slow growth that is an excellent choice for a formal townhouse garden (or any other use requiring a horticultural "pom-pom"). 'Sunnyside' has probably the largest leaves of the littleleaf boxwood forms—close to the size of the leaves of _Buxus sempervirens_—which gives it a unique texture for this group, but it suffers from winter bronzing if planted in full sun. 'Wintergreen' is an especially hardy form that will grow well as far north as zone 4. It has emerald-green foliage and a very uniform habit that makes it a good choice for a low formal hedge.

Two botanical forms of littleleaf boxwood are also frequently available. _Buxus microphylla_ var. _koreana_, Korean box, is an exceptionally hardy form (to zone 4) reaching 2 to 2.5 feet (0.6–0.76 m). Its leaves are rolled inward somewhat at the margins. _Buxus microphylla_ var. _japonica_, Japanese box, is particularly well adapted to areas with hot, wet summers. It tolerates heat well and shows some resistance to root rot and nematodes.

Recommended cultivars of common boxwood are numerous. 'Aureo-variegata' has yellow-variegated foliage. 'Bullata' eventually reaches 8 feet (2.4 m) in height and has interesting blunt, shortened leaves. 'Elegantissima' (Plate 166) is a slow-growing form with striking cream-variegated foliage (Plate 167); it is a spectacular specimen plant with time but is not the best choice for other uses. 'Glauca' has a blue bloom to the foliage. 'Graham Blandy' shows an extremely narrow and columnar habit; it has beautiful gray-green foliage. Two plants of 'Graham Blandy' flank the visitors' center entrance at the North Carolina State University Arboretum. 'Myrtifolia' has very small, neat foliage. 'Pendula' displays moderate weeping character. 'Pullman', with rich green foliage, boasts rapid, vigorous growth and exceptional cold-hardiness. 'Pyramidalis' has a handsome, narrowly pyramidal shape. 'Welleri' has a broad, rounded habit and resists winter burn.

These cultivars are only a tiny handful of the excellent cultivars available in the trade. For many years, the standard for "edging and hedges" was *Buxus sempervirens* 'Suffructicosa', English box. Wonderful old plantings of English boxwood can be found on the grounds of the Lincoln Memorial in Washington, D.C, for example. But many other cultivars offer better hardiness, better resistance to disease and insects, and unique color or form. They deserve greater attention in the landscape, especially for use as the foundation for garden structure.

At the North Carolina State University Arboretum, an extensive planting of boxwood selections, both young and more mature, adds quiet structure to the Southall garden adjacent to the brick house. Boxwood is remarkably versatile. It can be used in low hedges or as a small mass to delineate walls and corners, or planted in intricate knots and mazes. It is well suited for planting boxes, and it makes an excellent foil for early bulbs. The useful boxwoods are a practical but also beautiful component of a garden's foundation, a foundation made even more beautiful by the dusting of winter's first snow.

Camellia oleifera / Tea-oil camellia   PLATE 168

Camellia oleifera, the tea-oil camellia, is more cold-hardy than other camellias, thriving where winter temperatures sink to −10°F (−21°C). This attractive evergreen shrub flowers in fall and early winter, producing white or pinkish 2-inch (5 cm) blooms with a cluster of bright yellow stamens in the center of the flower. The charming flowers stand out against the glossy, emerald foliage like tiny white-and-gold porcelain teacups.

A native of China, this attractive shrub resembles the informal sasanqua camellia (*Camellia sasanqua*). It will reach 10–15 feet (3.1–4.5 m) in height with somewhat irregular habit of densely branched, multiple trunks in an overall broad pyramid that is less formal than the commonly grown *Camellia japonica*. The oval, glossy, dark green leaves are 2–4 inches long (5–10.1 cm). Tea-oil camellia is less likely than other, more tender camellias to suffer frost burn on the flowers. In Asia, the seeds are pressed for a source of cooking oil (*oleifera* means "oil-bearing").

Like all camellias, *Camellia oleifera* prefers acid, moist but well drained soil, with high organic matter. Camellias have shallow root systems and don't do well if planted too deeply, so be sure not to set the plant deeper than it grew at the nursery when transplanting. Because of the shallow roots, it is important to water camellias frequently during dry periods. *Camellia oleifera* is propagated from fresh seed, planted promptly, and from softwood cuttings, which are taken in the summer and rooted under mist.

The camellia, in its many variations, is a signature plant of the Southern garden.

The unusually cold-hardy *Camellia oleifera* extends the charm of these lovely shrubs to winter gardens in colder climates. White and gold blossoms will appear in the fall as if from nowhere, and their spell will resist winter's frosts. Camellias are spectacular in a walled garden, where the foliage and flowers are especially noticeable. They make a dramatic flowering hedge or evergreen screen and are a great choice to mix with hollies to add flowering interest to any evergreen mass.

Camellia spp. / Hardy camellia hybrids   PLATE 169

In winter, broad-leaved evergreens are an important part of the landscape. Many of our best-loved broad-leaved evergreens, such as hollies, are quite formal in habit, so any evergreen with more relaxed character is especially welcome, and even more so if it offers the added appeal of showy fall flowers to ease the transition into winter. The irresistible blossoms of camellias have beguiled gardeners in mild climates for centuries. These large shrubs or small trees with broad, evergreen leaves bear flowers in colors from white through pinks to deep red, of simple single form or of lush double, peony-, and rose-flowered forms.

Lack of cold-hardiness has restricted camellia culture to the southern half of the United States and to a narrow belt along the coasts. Hybridization programs have focused heavily on improving hardiness, especially after the exceptionally cold winters of the early 1980s, when many camellias were killed by unusually cold temperatures in areas where they normally flourished. By cross-breeding sasanqua camellia (*Camellia sasanqua*) and common camellia (*Camellia japonica*), both reliably hardy only to zone 7, with the much more cold-tolerant tea-oil camellia (*Camellia oleifera*) and other lesser-known hardy species and hybrids, breeders have achieved excellent hardiness in a new set of hybrids. As an added bonus, many of the new cultivars offer wonderful fragrance.

Close to 100 species of the genus *Camellia* exist in its native Asia, but only a few species are usually grown in the United States. *Camellia japonica*, the common camellia, blooms in late winter and early spring with large, deeply colored blossoms. The shrub has a dense, slightly more coarse and formal habit than other camellias. *Camellia sasanqua*, sasanqua camellia, blooms in the late fall and early winter with somewhat smaller flowers and a more refined and open habit than that of *C. japonica*. *Camellia sinensis*, the tea plant, blooms in early fall with small white flowers and emerald-green foliage, the source of tea buds and leaves. Many of the hardiest camellias are species that have never become common in American gardens, such as *Camellia oleifera*, the tea-oil camellia, which tolerates cold as low as −10°F (−21°C). Private and university breeders, as well as those at botanic gardens and arboreta, have been involved in the efforts to develop cold-hardy camellia hybrids. Some excellent cultivars have recently been produced by the breeding program of Dr. William Ackerman at the United States National Arboretum.

These hardy hybrid cultivars are selections from Ackerman's own ongoing breeding program. The parentage of the cultivars is relatively complex, including backcrosses to the hardy parents, and use of *Camellia sasanqua*, *C. oleifera*, and additional hybrids such as *C.* × *hiemalis* and *C.* × *williamsii*, as parents. The named selections are all similar to both *C. oleifera* and *C. sasanqua* in habit and are late-blooming, from late fall into early winter, with light blooming all the way into the new year, like their parents. They are all reportedly reliable through zone 6, but benefit from wind protection and light shade in zone 6.

Among those available to gardeners are 'Polar Ice', with snow-white single flowers that bloom quite late; 'Snow Flurry', with white flowers that bloom especially early; 'Winter's Charm', with early pink flowers and an upright habit; 'Winter's Dream', with pink, semidouble flowers; 'Winter's Hope', with white, semidouble flowers an open, spreading habit'; 'Winter's Rose', with pink, miniature double flowers that bloom heavily and early, and a compact, slow-growing habit; 'Winter's Star', with red-pink single flowers that sometimes have white centers; and 'Winter's Waterlily', with white double flowers that bloom quite late, and an upright habit.

Those who are especially interested in keeping up with new developments in camellia selection should consider joining the American Camellia Society, P.O. Box 1217-FL, Fort Valley, Georgia, 31030-1217.

Camellia culture is similar to that of azaleas and rhododendrons (*Rhododendron* spp.), in that they thrive in moist, well-drained, slightly acid soils in partial shade. Surprisingly, many camellias will do well in full summer sun after a period of acclimation and will probably flower more prolifically in full sun. However, foliage scorch (yellowing and bronzing of the foliage) can occur in winter when leaves are exposed to bright sun on cold days. Scorch can be prevented by siting plants where they receive more shade in January and February. This can be achieved without losing full summer sun by taking advantage of the lower angle of the sun in winter.

Camellias can be propagated from cuttings or seed, or by grafting. Softwood cuttings are rooted under mist. Seed requires no pre-treatment. Grafting desirable scion wood onto large root stock speeds early growth and promotes early flowering of young plants.

The new hybrids discussed here are hardy in areas where camellias are commonly grown now (zones 7–9), and they should be reliably cold-hardy into an extended range from traditional camellia culture areas, perhaps as far north as Philadelphia and coastal New England (zone 6).

The informal evergreen character of these camellias, combined with luscious fall and winter bloom, make them treasures in the winter garden. In a walled garden, against a severe wall, as a hedge, as a specimen near your favorite window seat in the sunroom, or just outside the front door, these hardy camellias delight the senses with the surprising large, bright blooms through the fall and early winter—what a happy holiday they make!

Cephalotaxus harringtonia / Japanese plum yew   PLATE 170

In winter, the qualities of evergreen foliage become especially important in the garden, and plants that combine this beautiful foliage with appealing form are at a premium. A star of the winter garden, *Cephalotaxus harringtonia*, Japanese plum yew, displays rich black-green foliage combined with graceful, elegant form. Japanese plum yew superficially resembles its close relative, *Taxus*, common yew, but it is a more refined plant, with rounded outline and long, slender, sharply tapered needles of a striking, uniform black-green. The plant brings a distinctive evergreen formality to the garden and yet requires little maintenance.

A native of Japan, *Cephalotaxus harringtonia* thrives in the kind of conditions that are anathema to most conifers. This marvelous conifer grows well in shade, tolerates wet soils, and flourishes in hot, humid conditions. Even the worst soils don't stop the growth of plum yew, which maintains excellent foliage color and quality under conditions that would kill many other conifers.

Japanese plum yew is a coniferous evergreen shrub (or, rarely, small tree). The species generally reaches 5–10 feet (1.5–3.1 m) in height. Ultimate habit and spread depend on the cultivar and range from prostrate to upright and bushy. *Cephalotaxus harringtonia* is a slow-growing plant, but this trait can be advantageous in gardens with limited space. Japanese plum yew transplants readily. It prefers moist, well-drained soil, but once it is established, it will tolerate a wide range of soils, from dry and sandy to heavy clays. Japanese plum yew is cold-hardy through zone 6 and in the warmer parts of zone 5. Japanese plum yew is propagated readily from cuttings, but it is a bit slow to root and grow on from the rooted cutting. Rooting promoters are applied to the cuttings, which are then rooted under mist.

Like common yew, Japanese plum yew is a conifer that does not bear cones, but instead, unlike common yew, it produces fruits that resemble a primitive olive. The male and female flowers are produced on separate plants; both must be present for the females to bear fruit. The seeds and foliage of all of the plants in this botanical group are generally quite poisonous if ingested but are not dangerous to handle.

Besides heat- and shade-tolerance, another advantage of *Cephalotaxus harringtonia* over *Taxus* species is that deer do not like the foliage of Japanese plum yew. Deer can completely defoliate a well-grown common yew in one feeding, which usually leads to the death of the hapless yew plant. Amazingly, the deer will not generally feed on Japanese plum yew in the same garden.

'Fastigiata' is an upright cultivar with densely columnar branching. Its needles are somewhat shorter than other *Cephalotaxus,* being more like true yew, and are whorled around the stem, giving an interesting textural effect. 'Fastigiata' is very popular in small landscapes. It makes an elegant low hedge or can be used as a specimen for patio planters and other confined spaces. It will eventually reach 8–10 feet (2.4–3.1 m) in height, developing a much more rounded habit as a mature specimen than it maintains as a young plant, but its slow growth makes this an extremely long-term process. At the North Carolina State University Arboretum in Raleigh, North Carolina, a magnificent 25-year-old specimen of 'Fastigiata' has reached 10 feet (3.1 m) in height with a spread of about 7 feet (2.1 m).

One of the best known cultivars is 'Duke Gardens', named for the garden of origin of this cultivar. Sarah P. Duke Gardens at Duke University in Durham, North Carolina, has some lovely specimens of this beautiful cultivar, a low-growing type with upswept arching branches. 'Duke Gardens' will eventually reach 4–5 feet (1.2–1.5 m) in height, but it is slow to gain height. It spreads in a lovely, semiformal, oval to rounded shape. 'Duke Gardens' is one of the few *Cephalotaxus* that is difficult to propagate from rooted cuttings. Surprisingly, the upright 'Fastigiata' is the parent plant of 'Duke Gardens'.

'Fritz Huber', another interesting cultivar, is a very low-growing form with dense foliage of exceptional quality. The unusual 'Korean Gold' displays gold new growth and has an upright habit similar to that of 'Fastigiata'.

Among the most widely available and useful forms of plum yew are the low-growing, spreading, prostrate forms. Their graceful undulating silhouettes make excellent groundcovers in shaded sites where other groundcovers are difficult to establish. The spreading forms have been confused over time, with a mix-up in names occurring as plants were produced. You will encounter low-spreading forms variously called *Cephalotaxus harringtonia* 'Prostrata', *C. harringtonia* var. *prostrata, C. harringtonia* var. *drupacea,* and *C. harringtonia* 'Nana'. Of them, the only name with any consistent utility is *C. harringtonia* 'Prostrata'. The best of them is a male clone, originally from Hillier Nurseries in England, which performs well in full sun or shade. 'Prostrata' recently received a Pennsylvania Horticultural Society Gold Medal Award for being an outstanding garden plant. A good example of this true *C. harringtonia* 'Prostrata' can be seen at the Brooklyn Botanic Garden. *Cephalotaxus harringtonia* 'Prostrata' is a beautiful informal, low, spreading plant with arching, slightly pendulous branches that arch up and out from the clump like a low fountain. The fluid effect is stunning, especially in winter when there are so many stark silhouettes.

Other species of *Cephalotaxus* may be found in botanical gardens and arboreta. One of the most beautiful is *Cephalotaxus fortunei*, a native of China with exceptionally long, slender needles and an upright, somewhat open habit. You may also encounter *C. koreana* and *C. sinensis*, from Korea and China respectively. Both are upright and shrubby, with needles intermediate in character between *C. harringtonia* and *C. fortunei*.

In addition to the older specimen near the visitors' center of the North Carolina State University Arboretum, a number of different *Cephalotaxus* can be seen in the Arboretum's conifer collections and in the shade house. The versatile nature of Japanese plum yew in the landscape and its dependably elegant beauty deserve the attention of all gardeners. Because of its unique foliage and habit, Japanese plum yew is one of the most beautiful of the needle-foliaged evergreens. It is an exquisite evergreen for shady sites and challenging soils. It is a spectacular foil for winter-blooming shade plants, such as mahonia. In the late winter light, the rich dark foliage adds luster and depth to the garden.

Chamaecyparis nootkatensis / Alaska cedar   PLATE 171

Conifers are stars in the winter landscape, but some gardeners fall into predictable plantings, avoiding unfamiliar conifers that have not yet received the stamp of approval from the neighbors. Many out-of-the-ordinary conifers are surprisingly adaptable. The North Carolina State University Arboretum has been a center for advocacy of new and unusual plants, including conifers, since J. C. Raulston began the Arboretum in 1976. Many of the conifers planted at the Arboretum for evaluation are conifers that "should not" or "could not" perform well in the hot, humid Southeast, according to plant dogma. But, surprisingly, many of these "doomed" conifers have turned out to be some of the best for the southeastern United States Piedmont, with its heavy clay soil and swings from drought to flood and back again.

Of the myriad of conifers that can (and should!) be used in more gardens, the Alaska cedar is a standout. This native of coastal Alaska, Washington, and Oregon is one of the most graceful conifers in nature and in the landscape. The foliage is a dark, rich, bluish gray-green and is draped in flattened sprays from drooping branches. The top of the tree ascends in a narrow, feathery spire above the uplifted arms of the branches, creating a distinctive and lovely profile, particularly when grown where it can be viewed against the setting sun or the winter tracery of deciduous trees. An informally conical tree, it can reach nearly 100 feet (31 m) in the wild but is generally smaller in the landscape, approaching 20–40 feet (6.2–12.4 m) with age.

The growth rate of Alaska cedar is strictly moderate—a surprising trait for one of the parents of the bionically fast-growing intergeneric hybrid Leyland cypress, × *Cupressocyparis leylandii* (whose other parent is *Cupressus macrocarpa*, Monterey Cypress). Alaska cedar makes an excellent tall specimen or screening plant, but it takes some time to reach full height. The advantage to its slower growth is that there is generally less tendency for the conifer to blow over or to "break up" (i.e., have limbs stretch down and away to separate from the main plant body).

Alaska cedar does best in evenly moist, well-drained soils in full sun or light shade.

It is a well-known landscape specimen in northerly parts of North America, but it has also performed miraculously in clay soils of the Southeast. It is fully cold-hardy through zone 4. Although native to northwestern coastal regions, Alaska cedar does not take kindly to the extremes of southeastern coastal summers; however, it performs well in a range of other climates from New England to the Midwest to the Deep South. It shows no pest or disease problems and maintains high-quality foliage year-round. Propagation of the species is from seed. Cultivars can be successfully rooted but are generally grafted for commercial production.

A few cultivars are available, primarily from the Northwest, where Alaska cedar is more widely grown and planted, but increasingly from nurseries on the eastern side of the United States as well. 'Pendula', the most readily available, has exaggerated weeping branches and an almost ghostly appearance. A newer cultivar is an upright, narrow form called 'Green Arrow', from Buchholz and Buchholz Nursery, with lighter green foliage than the species. Variegated forms include 'Variegata' and 'Laura Aurora', both with good splashes of cream throughout the foliage. 'Glauca' has unusually blue foliage; it is also available in a more weeping silhouette as 'Glauca Pendula'. 'Compacta' is a rounded, compact form. The cultivars are all well worth hunting.

Alaska cedar is best used in well-spaced groups or as a single, thoughtfully placed plant, so its magnificent form and foliage can be appreciated. Because of its moderate growth rate, the adaptable Alaska cedar also is ideal as a container or tub plant for patios and courtyards. There is a wonderful planting of the ghostly cultivar 'Pendula' on a slope at Pepsico headquarters in upstate New York. These corporate grounds are renowned for their collection of huge sculpture in a vast landscape, but to my mind, the Alaska cedars, floating down the hill like arboreal spirits, outshine many of the famous works of art.

Chamaecyparis nootkatensis is an exceptional conifer for the landscape. Its captivating form can transform a garden corner or knoll. At this time of the year when conifers are in special focus, Alaska cedar makes a picture of quiet drama and great beauty for gardens in winter.

Chamaecyparis pisifera 'Filifera Aurea' / Golden threadleaf sawara cypress
PLATE 172

It's the dead end of winter and everyone's tired of skinny brown trees. At this time of year, when we yearn for the first flash of daffodil yellow and the luscious magenta of early magnolias, who wants a low-key landscape? No one I know!

One plant that tops the list of affordable, desirable, ice-cream-sundae ornamentals is, believe it or not, an evergreen conifer—golden threadleaf sawara cypress, *Chamaecyparis pisifera* 'Filifera Aurea'. Even the name rolls around on your tongue like a sumptuous dessert—and sumptuous the plant is, some would say to the point of overload. Maybe it is a bit too much for folks whose garden wardrobes tend toward the drab, but how can you resist a plant that brazens a shamelessly bold, seamless yellow coat of drooping, thread-like foliage? In the winter garden, 'Filifera Aurea' positively glows as it waits for spring's brights to catch up with it.

Chamaecyparis pisifera is one of those easily propagated species that horticultural humanity has fooled with for a very long time. It roots readily from cuttings under mist, and the result is a plethora of cultivars. 'Filifera Aurea' is an old cultivar of *C. pisifera*, and no other selections of the species can approach this old reliable for pure shock value. The tightly compressed needles are wrapped around pendulous branchlets;

hence, the name "threadleaf." Another cultivar, 'Filifera', offers the same type of foliage in a nice quiet green. But 'Filifera Aurea' stores some of its excess sugars and starches that form in its needles as a bright gold pigment—a glorious, extravagant color that has to make any gardener smile, especially in the midst of the late winter doldrums.

'Filifera Aurea' is a good-sized plant, eventually reaching 18–20 feet (5.6–6.2 m) with age, but it retains a relatively formal, broadly conical shape that adds to its bold statement in the landscape. Other, smaller golden-threadleaf type cultivars, such as 'Golden Mop' and 'Filifera Aurea Nana', offer equally bold foliage but, because of their size, they're on the bench when it comes to brazen impact.

Many of the *Chamaecyparis pisifera* cultivars perform well in a widely diverse range of conditions. They will thrive in heavy clays, dry sands, and moist loams. 'Filifera Aurea' is no exception. It will happily grow a foot a year (if not overly stressed) in a broad range of soil conditions as long as it has full sun. It is hardy to zone 4. The cultivar is readily available at many nurseries and garden centers. Propagation of golden threadleaf Sawara cypress is dependable if cuttings with some hardened wood are taken in the winter months and rooted under mist.

Full sun is important for the plant to develop intense color. Parts of the plant that are in shade or facing to the north generally show less color and may even appear green in extreme situations. This is because lower light results in less photosynthesis and, therefore, less sugar and starch available to make that all-important pigment. For this reason, it is important to place 'Filifera Aurea' so that the south side is the one viewed. For example, if planting the cultivar along a path that runs east–west, place 'Filifera Aurea' on the north side of the walk, so that as you pass, you see the plant's south side. If you can't work this out on paper, go for a visit to the conifer collection at the North Carolina State University Arboretum. A beacon plant of 'Filifera Aurea' basks in the sun just east of the Arboretum's rose garden, on the north side of the broad path. This plant's southern side faces the viewer and is dramatically gold, while the plant's north side faces north and is significantly less brilliant with a much greener tint to the foliage. But the side that you'll notice and never forget is the bright southern side.

Chamaecyparis pisifera 'Filifera Aurea' seems to be the kind of plant that inspires great feeling. It has been called "ghastly," "tacky," and "a horror"—but it is not. It is a plant with one mission in life—to bring out the playful side of gardens and to brighten up corners, not by relying on subtlety but with a blast of color during the worst of the winter doldrums. It can brighten your garden in many ways: as a central focus for a planting of other mixed evergreens, as a splash of color alone in a corner, as a bright surprise around a bend in the path, or as a beacon near the top of a rise. In fact, *Chamaecyparis pisifera* 'Filifera Aurea' has a place in almost every garden. Just try to walk past this playful plant in the dreary end of winter without getting the urge to give it a hug.

Chimonanthus praecox / Fragrant wintersweet   PLATE 173

Just the visual pleasure of a flower's form and color is precious enough in the winter garden, and a plant that brings not only flowers but fragrance to the winter landscape is a special prize indeed. Fragrant wintersweet, *Chimonanthus praecox*, is just such a treasure. Starting in December or January in zone 7, and continuing through the winter, translucent, waxy yellow flowers open along the bare stems to reveal purple centers, and send a heavenly scent singing through the garden. The lemony, spicy aroma is distinctive, without being too heavy or overbearing. Just one vigorous shrub lightly and sweetly scents the entire northeastern corner of the North Carolina State University Arboretum.

Native to China, *Chimonanthus praecox* was introduced to cultivation in 1766. *Chimonanthus* is Greek for "winter flower," and *praecox* is related to the term "precocious." The common name for this plant, "fragrant wintersweet," is just as appropriate a name as the Greek.

Wintersweet is a loose, upright, deciduous shrub that can reach 10–15 feet (3.1–4.5 m) tall and spread to 10 feet (3.1 m). In addition to the fragrant flowers in winter, wintersweet has broad, emerald-green foliage through the summer which turns a variable but attractive yellow in the fall, adding seasonal interest to its own corner of the garden. The branches of *Chimonanthus praecox* are excellent to cut for indoor forcing, as the blooms are lovely and will bring a refreshing fragrance to any room.

Fragrant wintersweet is easily transplanted and will take to many soils as long as the site is moderately well drained. It needs full sun for best performance. *Chimonanthus praecox* is reliably hardy to zone 6 in terms of survival, but it will flower best in zone 7 and warmer. A protected site will help maximize flowering in all zones. Pruning out old canes after flowering will insure vigorous growth and flowering again the following year. Wintersweet is propagated both from seed and cuttings. The seed is collected in early summer, before it begins to dry. Cuttings are taken in July.

In your own garden, plant *Chimonanthus praecox* near a path, gate, porch, or anywhere you will be likely to pass, so that you can enjoy the light, sweet scent. It is wonderful near your entryway door as a fragrant welcome home, or next to your favorite sheltered bench in the garden where it can entice you to sit for a bit in the late winter sun.

Cornus alba and *C. sericea* / Tatarian and red-osier dogwoods   PLATES 174, 175

What a change in the landscape winter brings—and what a wonderful opportunity to garden with striking color and form. Two plants with especially vivid winter color are the shrubby dogwoods: *Cornus alba*, tatarian dogwood, with its brilliant red or yellow stems; and the vivid red-twigged *C. sericea*, red-osier dogwood. The bold stems in winter add strong colors to the landscape when most plants are garbed in muted tones of brown and gray.

These two dogwoods are often confused: both species have a number of cultivars with red-colored stems; however, only *Cornus sericea* has cultivars with bright yellow stems. Both of these winter-interest shrubs bear clusters of small, whitish flowers in early summer, which mature into bluish-white fruit. The green summer foliage can turn to beautiful shades of purple and red in the fall, but this is a very variable trait, and the winter bark is generally their showiest trait.

Cornus sericea (Plate 174) is native to eastern North America, including most of Canada and reaching south to Kentucky. It can be seen growing in low, wet areas of woods and abandoned fields. It grows well from zone 9 to zone 2. Red-osier dogwood will reach 5–9 feet (1.5–2.7 m) in height with an equal or greater spread. An older botanical name for this plant was *C. stolonifera*, referring to its underground spreading stolons. The plant will spread into a mounded colony of many upright stems with short, horizontal branching. It can be a bit invasive, as it is a vigorous plant that grows rapidly; however, this can be advantageous in difficult sites. It is well adapted to a broad range of soil conditions in the landscape and will tolerate everything from swampy to dry conditions. Red-osier dogwood is most striking in full sun where stem color is best developed and most easily seen. It is easily propagated from rooted cuttings taken anytime during the year when leaves are on the stems, or from seed layered in moist peat and held at 40°F (4°C) for 2 to 3 months.

There are several red-twigged cultivars of *Cornus sericea*, but perhaps the most well known and readily available cultivar of this plant is 'Flaviramea' (sometimes seen as 'Lutea'), yellow-twig dogwood (Plate 175). The stems of this cultivar are a vibrant canary-yellow and glow like lasers in the winter landscape. They are especially beautiful when grown near a red-twigged cultivar, or in front of a dark evergreen holly or mahonia. As with all of the colored-stem dogwoods, care is needed to avoid misplacing this bold beauty and drowning out other, quieter garden voices. 'Flaviramea' only works well as a large mass on large properties. On small sites, an individual plant can be a welcome note of color in a perennial border or mixed shrub border. It is lovely against dark evergreen foliage.

Recommended red-twigged cultivars of *Cornus sericea* include 'Cardinal' and 'Isanti', both introductions from the Minnesota Landscape Arboretum. There are also cultivars with variegated foliage, which add interest during the summer. One excellent variegated cultivar is 'Silver and Gold', an introduction from the Mt. Cuba Center for Research on Native Piedmont Plants, in Pennsylvania. It was selected from a mutation found on a plant of 'Flaviramea'. 'Silver and Gold' has bright yellow stems and a creamy white margin on the leaves.

Cornus alba, the tatarian dogwood, is a native of Siberia and Korea. There are no yellow-stemmed cultivars of this dogwood, but there are many red-stemmed cultivars, several of which exhibit a range of interesting foliar variegations during the spring and summer. Tatarian dogwood is very similar in size and habit to red-osier dogwood, although not quite as potentially invasive. Nonetheless, it is very vigorous and can still take over an area from other less rampant plants. *Cornus alba* is also adapted to a range of soils but prefers moist, well-drained sites in full sun. The best stem color develops on young, vigorous growth. These young shoots are encouraged by cutting back older and larger branches to 6–12 inches (15.2–30.4 cm) in late winter each year. This pruning will also help to control its spread. *Cornus alba* is somewhat more cold-hardy than *C. sericea*, but it is significantly less well-adapted to regions with hot, humid summers. *Cornus alba* is propagated like *C. sericea*, but cuttings can be taken at any time of year and root readily under mist.

The soft gray-green leaves of the *Cornus alba* cultivar 'Argenteo-marginata' (or 'Elegantissima') have creamy margins; the foliage is lovely against the bright red stems, even in summer. 'Sibirica' has carmine stems and bluish fruit and may be somewhat less invasive than others. 'Spaethii' has leaves with bold yellow borders that hold their color through the summer season better than other *C. alba* cultivars, in which the variegation may fade to green in the heat of summer.

Spears of scarlet and gold stems are arresting in a winter garden, whether they are piercing a blanket of white snow or framed by the black-green of a holly hedge. The touch of color from these remarkable shrubs goes far to invigorate a winter day. In the mixed border or in a wild area, red-osier and tatarian dogwoods punctuate the winter garden with bold, vertical strokes of red and yellow.

Corylus avellana 'Contorta' / Harry Lauder's walking stick   PLATE 176

There just is no escaping it—eventually, at some level or another, all gardeners succumb to the quest for the rare and unusual. This yen may manifest as the drive to find and rescue the rarest of native populations of a tiny fern with only five remaining plants that grow in only one spot on the entire planet (currently endangered, of course, by planned construction of a trans-global shopping mall) or it may develop as an insatiable hunger for a cutting of that dwarf, contorted, pink-and-gold-variegated, cut-leaved, sterile, chartreuse-flowered form of a hitherto-believed-to-be-extinct, cold-hardy to zone 1, heat-tolerant, broadleaved evergreen shrub rumored to now exist only in the collections of the extremely remote Atlantis Botanic Garden (a garden known only to a few seriously intrepid collectors, which refuses to participate in Index Semina exchanges).

Whatever form this yearning for the unusual takes, even the most blasé of horticulturists eventually find themselves searching for choice plants of one form or another. One magnificent plant that has long been a traditional source of choice garden character is *Corylus avellana* 'Contorta', Harry Lauder's walking stick. This unusual shrub or small tree is a contorted form of the commercial European filbert, *Corylus avellana*, which is grown and highly valued for its delicious nuts. The branching of this form is twisted into striking, spiral contortions throughout the entire plant. It is a spectacular addition to the winter garden, where the sculptural patterns created by the branches can be clearly seen.

Native to Europe and parts of Asia and northern Africa, the species *Corylus avellana* is a small tree or large, woody, multistemmed, thicket-forming shrub. Its deciduous dark green foliage, about 3–4 inches (7.6–10.1 cm) long and almost as wide, is rather coarse and hairy. The flowers are tiny, with the male flowers borne on long, narrow catkins, and female flowers in shorter, thicker catkins, similar to those of its close relatives the birches (*Betula* spp.) and alders (*Alnus* spp.) The catkins of male flowers are yellow and put on a handsome show in late winter before the leaves emerge. The female flowers are much more subtle and require closer inspection to see the delicate but amazingly carmine-colored floral parts emerging from the buds.

Corylus avellana 'Contorta' is much like its parent species with one important exception—its contorted growth. This fascinating plant is interesting in summer because the leaves are also somewhat contorted, but it is at its peak in late winter, when the catkins of yellow male flowers dangle from the twisted branches.

Harry Lauder's walking stick was discovered in a hedgerow in England in the mid-1800s and propagated for its unique habit. It flowers somewhat later than the species, in late winter and early spring. It does not bear fruit. The plant will reach 10 feet (3.1 m) with some age, but it is a relatively slow grower, which makes it an excellent specimen plant for small gardens. It is completely hardy to zone 4.

Corylus avellana 'Contorta' will perform well in a range of soils in full sun or with a little shade. It is propagated by grafting scion wood of the cultivar onto rootstock of the species. The species understock does best in deep, loamy soils but will also perform well in clay soils. The understock tends to sucker and the suckers must be continually removed to avoid overgrowth of the cultivar.

Harry Lauder's walking stick is an attainable, impeccable beginning in the search for choice, unusual plants for the garden. At the North Carolina State University Arboretum, this intriguing plant beguiles visitors at the entrance arbor where it is a harbinger of the exciting collections of choice and even more rare plants to be found in the gardens inside.

Cryptomeria japonica / Japanese cedar   PLATE 177

Japanese gardens around the world are places of repose, visual elegance, and welcome quiet, offering a restful escape from our hectic modern world. The Japanese-style garden can be especially enticing in winter when the sculptural beauty of its conifers and its artfully placed stones and paths are not at all camouflaged by deciduous foliage and flowers.

One conifer that can be seen in many Japanese-style gardens, including that of the North Carolina State University Arboretum, is the Japanese cedar, *Cryptomeria japonica*. As the name suggests, this tree was introduced from Japan in the 1800s. Japanese cedar is a beautiful evergreen conifer with a uniquely refined texture that is both soft and crisp in the landscape. Its needles are very different from those of the more familiar pines and spruces. They are awl-shaped and arranged spirally around each branchlet. Each needle arches up toward the tip of its branch, making the individual needles distinctly visible. From a distance, this thickens the appearance of the branch and actually softens the texture of the plant, while at close hand, the branches appear to be covered with tiny curving spikelets which, contrary to their appearance, are smooth to the touch and very "gardener-friendly." Japanese cedar has a fascinating texture, both as a specimen and in a group planting.

Cryptomeria japonica grows in a pyramidal, semiformal shape and can reach heights of 50–80 feet (15.2–24.8 m). It prefers a rich, deep, acid soil but will thrive in a range of soils from sands to clays. Like almost all conifers, it needs full sun to grow well but it will take light, high shade. It is generally reliably hardy to zone 6, with a few exceptional cultivars (e.g., 'Yoshino') possibly surviving in protected areas to zone 5.

Japanese cedar is one of the best evergreen conifers for gardens because of its great beauty and tough adaptability to a range of climates. Seedling trees of this species have been planted since the 1800s. Unfortunately, the seedling trees often age poorly, becoming ratty and open. Planting named cultivars will avoid this problem and allow better use of this excellent conifer.

Many horticulturists are familiar with the full-size, rapidly growing forms of this plant, 'Yoshino' and 'Benjamin Franklin', which are becoming popular choices for screening, especially as alternatives to Leyland cypress (× *Cupressocyparis leylandii*). 'Yoshino' will reach 50 feet (15.2 m) quite rapidly and retain a uniform, informally pyramidal habit with a slightly irregular, cloudlike silhouette. It is the most reliably cold-hardy and the best choice for zone 6 gardens. 'Benjamin Franklin' is equally fast but has slightly blue-green foliage. It was selected for tolerance to salt spray and wind.

Many horticulturists are not as aware, however, that there are over 70 cultivars of Japanese cedar with an incredible diversity of form, color, mature size, and growth rate. Cultivars range from tiny dwarf plants to tall trees, with many variegated and unusual foliage forms available. In fact, there is a perfect cultivar of Japanese cedar for each and every garden in zones 6 and warmer.

The primary reason for this great abundance of forms of *Cryptomeria* is because this species is revered in its native Japan both as an important forest tree and as a beautiful ornamental. Japanese cedar has been the subject of horticultural attention and selection for hundreds of years, not only in Japan, but also, to a lesser extent, in the gardens and nurseries of Europe and North America.

There are many cultivars of Japanese cedar with a wide range of horticultural characteristics to offer. For rapid, beautiful evergreen screening, 'Yoshino' and 'Benjamin Franklin' are excellent full-sized, dark green cultivars. Some of the other possibilities include extreme dwarf cultivars ('Tenzan-Yatsubusa'); forms with snakelike branches ('Araucarioides'); trees with fluffy juvenile foliage that turns purple or gold in winter ('Elegans', 'Elegans Aurea') (see the following entry for more on these cultivars); and variegated cultivars ('Sekkan-sugi' [Plate 177], 'Knaptonensis'). The myriad of Japanese cedar cultivars offers unbelievable variety in the landscape.

Japanese cedar can be readily propagated by cuttings, which are best taken from September through March (or whenever hardened wood is available). Cuttings are treated with rooting promoters before being rooted under mist. *Cryptomeria* can also be grown from seed, but this is difficult, and, of course, seed from a cultivar such as 'Yoshino' will not grow up to give you another 'Yoshino' plant. Asexual, or vegetative, propagation is the only way to exactly reproduce a cultivar. Named cultivars of Japanese cedar are preferable for all garden and landscape use.

In the quiet, wintry time of year, the mystique of a Japanese garden appeals to our sense of peace. The graceful Japanese cedar softly frames the west wall of the Japanese garden at the North Carolina State University Arboretum. Walk past the black bamboo, over the bridge of spirits, and let the unique, evergreen beauty of Japanese cedar inspire your thoughts with contemplative quietude.

Cryptomeria japonica 'Elegans' and 'Elegans Aurea' /
'Elegans' and 'Elegans Aurea' Japanese cedar, Plume cedar   PLATE 178

Two cultivars of Japanese cedar that are especially beautiful in the winter garden are 'Elegans' and the closely related 'Elegans Aurea'. Both are both a bright emerald-green in the spring, summer, and fall, but as soon as the weather turns cold for a few weeks, both cultivars begin a fascinating transformation. The foliage of 'Elegans' turns a deep burgundy-plum color, while that of 'Elegans Aurea' glows a startling bright lime-gold.

'Elegans' and 'Elegans Aurea' are medium-sized tree forms reaching 20 feet (6.2 m) and greater with age, with an informal, rounded, pyramidal habit. Both are slower growing than 'Yoshino' and 'Benjamin Franklin', the cultivars recommended for screening. 'Elegans' and 'Elegans Aurea' show growth rates on the order of 1 foot (0.3 m) a year, as opposed to the 3 feet (0.9 m) per year and greater that 'Yoshino' and 'Benjamin Franklin' are capable of.

'Elegans' and 'Elegans Aurea' have unique, completely juvenile foliage. The juvenile foliage of very young seedling conifers of all species is generally much softer, more feathery, and less distinctive than the adult foliage that will eventually develop on the plant as it matures. Cultivars of many species of conifers have been selected, however, for permanent retention of the juvenile foliage. 'Elegans' and 'Elegans Aurea' were selected, in part, for their wonderfully soft-textured juvenile foliage, which can be compared with the softness of goose down (even the most inveterate conifer-phobe will never be "stuck" by the feathery foliage of these two Japanese cedars). They are especially distinctive among juvenile-foliaged conifers because the foliage of each changes to a dramatically different color in the colder weather of winter.

When planted together, the rich plum of 'Elegans' combines with the blazing gold of 'Elegans Aurea' to create a study in garden color that defies description. These two Japanese cedars are fabulous companions for perennials, especially tall perennials, and are unique anchors for a border. They are a fresh green foundation in summer, while the perennials are blooming, and are transformed into bold sculptures of spectacular color while the perennials are dormant.

Because of their moderate size and growth rate, these two *Cryptomeria* cultivars are also ideal for use in combination with other specimen woody plants, which might be overshadowed and outcompeted by more robust growers. Their moderate growth rate also makes them excellent choices to bring the interest of an unusual specimen or the dazzle of an incredibly bold hedge to smaller landscapes.

Like other Japanese cedars, 'Elegans' and 'Elegans Aurea' are well adapted to heavy, wet clay soils, such as those of the Southeast. They are tolerant of most soils as long as they are not subject to long periods of drought. Hardy to zone 7, 'Elegans' and 'Elegans Aurea' are subject to foliar dieback when temperatures dip into the teens. These two cultivars can be grown in colder areas than zone 7 with winter protection or in a sheltered site, but beware of damage from late spring frosts on early flushes of growth in very sheltered sites where the plants may be fooled into pushing out early. Late frost-burn can be cut off and the plant will recover. It is especially important to keep these two cultivars out of very windy sites to avoid needle scorch.

Both 'Elegans' and 'Elegans Aurea' are particularly easy to root from cuttings because of their juvenile character. Cuttings can be successfully rooted any time of the year (except during the peak of the first spring flush of growth), as long as there is some wood on the cutting. Cuttings are treated with rooting hormone and rooted under mist. At the North Carolina State University Arboretum, *Cryptomeria japonica* 'Elegans' and 'Elegans Aurea' glow from the winter garden with vibrant hues. They are among the most beautiful trees available to light up our gardens in winter.

Danae racemosa / Poet's laurel   PLATE 179

Gardeners rely on conifers and broadleaved evergreens to add depth and definition to the winter landscape. In spite of their botanical diversity, these plants are often united in the minds of gardeners by their potential role in the winter garden. Poet's laurel, or Alexandrian laurel, is one of the most exquisitely elegant evergreens for the garden. It has a gracefully open habit, with slender branches that arch up and away from its crown like wings from the back of a swan. This small, evergreen shrub grows so slowly that it is rarely seen larger than 2–3 feet (0.6–0.9 m) in height in the landscape, with an equal spread, although it can reach greater size with time.

The "leaves" of poet's laurel are actually flattened stems, 2–3 inches (5–7.6 cm) long, which are technically referred to as *phylloclades*; the true leaves are tiny bractlike structures, barely visible, adjacent to the phylloclades. These leaf-like structures are among the most beautiful "foliage" in the plant world. Glossy and emerald-green, with a refined, tapered oval outline, each "leaf" is distinctly silhouetted and balanced gently away from the main stem, which terminates in yet another perfect "leaf." With some shade, the phylloclades look magnificent 12 months of the year, so glossy and richly colored that they look as if they were brushed into place by a Japanese master of painted silks. The branches of poet's laurel are wonderful for cut arrangements because they keep beautifully fresh-looking for long periods of time.

Danae racemosa is actually a woody member of the Liliaceae botanical family, although the insignificant, yellowish flowers of this shrub are hardly reminiscent of the great trumpets generally associated with lilies. The spring flowers ripen in the fall into a few delightful carmine berries, which dangle appealingly from the "leaf" axils. The bright, inedible fruit is about the size of wild cherries, 0.5 inch (1.2 cm) in diameter. It is a welcome visual treat as the season turns into winter.

Poet's laurel is native to western Asia. It needs at least part shade to thrive and does well in full shade. Full sun results in foliar scorch and plant decline over time. The shrub will grow in a range of soils as long as it is kept relatively moist. It is reliably hardy only through zone 8 but has performed well in protected microclimates in zone 7. *Danae racemosa* has no significant pest or disease problems, but it is very difficult to propagate, which is much of the reason why it is a difficult plant to find in the trade.

PLATE 162. The brilliant branches of *Acer palmatum* 'Sango-kaku' add a blaze of color to the winter garden. (WAYSIDE GARDENS)

PLATE 163. Coral-bark maple's scarlet twigs are a stunning contrast with the bright blue sky of a clear winter's day. (ROBERT HAYS)

PLATE 164. The silvery bark and architectural branching of European alder (*Alnus glutinosa*) are strong statements in the winter landscape. (KIM TRIPP)

PLATE 165. The winter catkins of many alders, including *Alnus firma*, add unusual interest to the winter garden. (KIM TRIPP)

PLATE 166. *Buxus sempervirens* 'Elegantissima' combines compact pleasing habit with brightly variegated foliage.

PLATE 167. *Buxus sempervirens* 'Elegantissima' foliage is variegated with broad margins of creamy white to gold.

PLATE 168. Hardy *Camellia oleifera* flowers prolifically with clear white flowers.

PLATE 169. *Camellia sasanqua* × *Camellia oleifera* is an exceptionally hardy hybrid with especially attractive flowers as well.

PLATE 170. *Cephalotaxus harringtonia* 'Prostrata' is an excellent choice for low masses and mounding groundcovers.

PLATE 171. *Chamaecyparis nootkatensis* 'Aurea' is a gold hued form of the useful and beautiful Alaska cedar.

PLATE 172. *Chamacyparis pisifera* 'Filifera Aurea' is an eye-popping addition to the landscape in winter and brings out the brilliance in early spring bulbs as well.

PLATE 173. *Chimonanthus praecox* flowers are fragrant and lovely.

PLATE 174. Red-twig dogwood stems stay bright red all winter.

PLATE 176. The contorted branching of Harry Lauder's walking stick (*Corylus avellana* 'Contorta') creates living sculpture in the winter landscape.

PLATE 175. Yellow-twig dogwood adds dramatic color to a winter garden.

PLATE 177. *Cryptomeria japonica* 'Sekkan-sugi' is a tall, strikingly variegated cultivar of Japanese cedar.

PLATE 178. This very old specimen of 'Elegans' Japanese cedar shows the plum winter color of its distinctive juvenile foliage.

PLATE 179. Poet's laurel (*Danae racemosa*) makes an elegant low-growing evergreen.

PLATE 180. The bright appeal of *Daphne mezereum*'s flowers is matched only by their fragrance.

PLATE 181. *Ficus pumila* is an unusually fine-textured evergreen vine.

PLATE 182. 'Bonfire' is only one of many deciduous hollies with brilliant red fruit that persist through the winter.

PLATE 183. As the fruit of *Ilex cornuta* 'D'or' ripen they change from light green to bright gold.

PLATE 184. The foliage of 'O'Spring' holly appears tie-dyed with bands of cream and gold.

PLATE 185. A cascade of winter jasmine's yellow blooms brightens the winter garden.

PLATE 186. 'Silver Mist' shore juniper is an unusually compact and silvery blue selection.

PLATE 187. 'Grey Owl' is a unique cultivar of eastern red cedar with a rounded, spreading habit and silver-gray foliage.

PLATE 188. Leatherleaf mahonia bears large clusters of purple blue fruit.

PLATE 189. The lemon-yellow flowers of *Mahonia* 'Arthur Menzies' are a cheerful addition to the late winter garden.

PLATE 190. The twisted branches of contorted mulberry (*Morus australis* 'Unryu') create a cluster of unicorn horns in the landscape.

PLATE 191. 'Fire Power' is one of the most popular cultivars of nandina because of its compact habit and dependably dramatic color.

PLATE 192. 'San Gabriel', with its delicate, finely cut foliage, is one of the most unusual cultivars of nandina.

PLATE 193. The vaselike silhouette and horizontally spreading branching of Amur cork tree are especially appealing in winter. (KIM TRIPP)

PLATE 194. The tall, narrow habit of *Picea omorika* 'Pendula', with its branches raised like gracefully curving arms, offers unforgettable garden character.

PLATE 195. 'Foxtail' blue spruce combines heat tolerance with excellent blue foliage and unusual foxtail-shaped branch outlines.

PLATE 196. Lacebark pine is well named for its opulent, many-hued bark.

PLATE 197. The red bark of Tanyosho pine is especially showy on multistemmed specimens.

PLATE 198. *Pinus densiflora* 'Oculus draconis' is one of the showiest variegated conifers available for the garden.

PLATE 199. The dragon's-eye Himalayan white pine offers more subtle, frosty variegation that combines well with many flowering shrubs and perennials. (KIM TRIPP)

PLATE 200. With their unusual shape and size, the dwarf loblolly pines at the NCSU Arboretum are a remarkable presence in the winter landscape. (KIM TRIPP)

PLATE 201. Hardy orange (*Poncirus trifoliata*) creates an impenetrable, thorny hedge or mass that is, nonetheless, pleasingly decorated with its small, orange fruits.

PLATE 202. The flowers of Japanese flowering apricot (*Prunus mume*) create a cloud of color against the winter sky.

PLATE 203. Japanese flowering apricot is a miracle of bloom in late winter.

PLATE 204. Japanese umbrella pine (*Sciadopitys verticillata*) has a reliably pyramidal habit that makes it an excellent specimen conifer.

PLATE 205. The unique needles of Japanese umbrella pine are flattened and arranged in whorls like the spines of an umbrella.

PLATE 206. The foliage sprays of 'Zebrina' western red cedar are banded with light gold in such a way as to appear zebra-striped from a distance.

Poet's laurel does not root well from cuttings, but parent plants can be successfully divided. Propagation from seed is a slow, tedious process. Berries must be collected, the seed separated from the flesh of the berry, and then the seed must be stored for as long as 1 to 2 years in a moist medium while the embryo develops. Once signs of germination are apparent, seed can be sown, but even then, seedlings take 2 to 3 years to reach a marketable size. This protracted production time is why it generally takes a good deal of hunting to find this remarkable plant in the trade.

Poet's laurel is a horticultural gift for the shaded areas of the winter garden. Its quintessential elegance and quiet refinement are more consistently rewarding than the flashy qualities of the showier broadleaved evergreens. The subtle beauty of this botanical gem offers landscape character found in few plants. It makes a unique drift of evergreen color in a rock garden or against a bamboo fence, where it adds Japanesque character. It is an effective companion for evergreen ferns and heaths and heathers and is especially handsome with the foliage and flowers of hellebores.

Daphne mezereum and *Daphne odora* / February daphne and Winter daphne   PLATE 180

"It's around that corner."

"No, I'm certain it's beyond the border."

"Oh, surely it's on the other side of that bench."

These are not the observations of a flock of birders in search of a rare warbler, nor are they even the exclamations of a grove of garden club members in search of a new nursery. They are the remarks of any number of late-winter visitors to the North Carolina State University Arboretum as they search for the source of one of the most deliciously enticing fragrances in the botanical world—the fragrance of winter daphne, *Daphne odora*.

Winter daphne is only one of a number of choice species belonging to the *Daphne* genus. Most of these fragrant plants bloom in early spring, but two of the species bloom in winter, perfuming the crisp air with a lush, lemony, floral scent that is irresistible. *Daphne odora*, winter daphne, and *D. mezereum*, February daphne, bloom anywhere from late January through early March with clusters of rosy pink to lavender blooms.

Daphne odora, winter daphne, is an evergreen, rounded shrub reaching 3–4 feet (0.9–1.2 m) in height with an equal spread and a very handsome, dense habit. In January, the flower buds swell into tight, colorful pink clusters. The buds open in February into charming, bell-like florets that persist into March.

Daphne mezereum, February daphne, is a semievergreen to deciduous shrub of somewhat larger and coarser habit than winter daphne. It has equally delightful flowers and fragrance (Plate 180) and the added bonus of attractive cherry-red fruits. The fragrance from the flowers is superb.

All daphnes, including *Daphne odora* and *D. mezereum*, have a reputation for being temperamental in the garden. This reputation is based on a strong aversion to being moved once established and on a certain propensity for the plant to die suddenly for little or no apparent reason. This is especially true of *D. mezereum*, which suffers from a sort of sudden-death virus. If transplanted with care, however, and sited correctly, both of the winter-blooming daphnes can give years of elegant garden delight.

Daphne mezereum and *D. odora* prefer moist, well-drained, neutral pH soil, but they will tolerate more acidic soils. *D. mezereum* is hardy to zone 4 but *D. odora* is reliable only to zone 7. Both do best in light shade. At the North Carolina State University Arboretum in Raleigh, North Carolina, both species have performed well in relatively heavy clay soils, but they have not been moved since they were first planted. Moving daphne after it has been planted is a sure way to cause sudden death in this plant.

A number of cultivars of the two winter-blooming species are also worthy of attention. There are three white-flowered forms of *Daphne mezereum*. One is a botanical form, *D. mezereum* var. *alba*, with creamy white flowers, which can be grown from seed, and two are cultivars. Both 'Paul's White' and 'Bowle's White' are cultivars of *Daphne mezereum* with pure white flowers. Winter daphne also has a white-flowered form, a cultivar called 'Alba', with cream-white flowers. 'Aureo-marginata' is an especially cold-hardy form of *Daphne odora*, with leaves lightly margined in gold. 'Mazellii' is an interesting form of *Daphne odora* that bears flowers in the axils of leaves as well as at the end of its branches. 'Variegata' is another yellow-variegated form of winter daphne, with brighter, more distinct marginal variegation and lighter pink flowers than 'Aureo-marginata'.

Daphne odora and *D. mezereum* can be propagated by rooted cuttings but they are somewhat inconsistent rooters. Cuttings with hardened wood are taken in early summer, treated with rooting promoters, and rooted under mist. Freshly harvested seed can also be sown and successfully germinated.

The enchanting fragrance of winter-blooming daphnes is an extraordinary presence in the garden, made even more so when either of these is planted near a walk or patio. However, there is also appeal to planting daphne in a slightly mysterious spot, placed to draw visitors to the garden on into the plantings in search of the center of that tantalizing aroma.

Ficus pumila / Climbing fig   PLATE 181

As winter's grip begins to loosen, but before the flush of spring color has arrived, a simple green blanket of leaves can seem a spectacular tapestry. While this same evergreen foliage cover may go unnoticed in the extravagance of summer, it is very valuable in late winter and often a neglected concept in garden planning. *Ficus pumila*, climbing fig, provides just such welcome relief in winter, and it is a handsome addition to the garden during the rest of the year as well.

Climbing fig is an evergreen twining vine, native to China and Japan, which holds fast to surfaces using aerial rootlets. It is similar to ivy (*Hedera* spp.) in that respect, but it is actually a botanical relative of mulberry (*Morus* spp.). As with *Hedera*, *Ficus pumila* can be found with two different leaf forms. The juvenile leaf form is more usually seen. The emerald-green juvenile foliage is small, about 1 inch (2.5 cm) long, and fine-textured. The leaves are somewhat heart-shaped, with a very short petiole attaching the leaf to the vine. The larger adult foliage is 2–4 inches (5–10.1 cm) long, leathery, dark green, and more oval in shape, with longer petioles. Juvenile foliage is produced initially until the plant reaches a certain stage in development. Adult foliage is produced when the vine has reached physiological maturity and is able to flower and bear fruit. This shift in "maturity" is not necessarily time related; it generally occurs after the vine has reached a certain height above the ground. Adult or juvenile character can be propagated vegetatively, which is why adult *Hedera* forms are available at nurseries and garden centers even though the adult plants for sale are not all 50 feet (15.2 m) tall.

Climbing fig is a relatively rapid grower and gracefully covers most surfaces with lovely interwoven ribbons of green. It thrives in high humidity and prefers a moist, well-drained soil but will tolerate clay soils as well. *Ficus pumila* will perform well in sun or partial shade, but does best with light shade. It is reliably hardy through zone 8, but temperatures in the range of 0°F (−16°C) will cause significant die-back. In areas where temperatures drop to single digits, the vine can still be grown as a woody perennial. If the vine is damaged by cold, it can be cut back in late winter and new stems will reemerge from the still-living roots. Climbing fig is an easily propagated plant that roots readily from cuttings taken any time of year and rooted under mist.

Cultivars of note include 'Minima', which has juvenile foliage with especially small leaves, and 'Variegata', which produces leaves with white markings. Another gardenworthy species, _Ficus nipponica_, is a recently introduced climbing fig from Korea that may have improved hardiness. The elongated, pointed oval foliage is a deep green and retains its beautiful quality throughout the year. A good specimen of _Ficus nipponica_ can be seen growing in the shade house at the North Carolina State University Arboretum. This is a special plant to watch for as it becomes available in the trade.

Late winter is an excellent time to consider where to plant climbing fig in your garden, because the bones of the landscape are bare and the best site for this wonderful climber is most easily determined. _Ficus pumila_ can turn a rough wooden surface or a bare stone wall into a delightful emerald tapestry, bringing life to the winter garden.

Ilex spp. / Deciduous hollies   PLATE 182

A brilliant pool of scarlet or gold is a delightful surprise when set among winter's somber weave of silvery gray branches and feathery conifers. The deciduous hollies are exactly right for such a display of startling, persistent color.

Deciduous hollies are a group of _Ilex_ species and hybrids closely related to the familiar evergreen hollies, but they lose their leaves each fall. Far from being a disadvantage, this annual leaf-fall allows for an even better show of the intensely colored red or gold berries left behind on the branches. While small creamy white or greenish flowers are borne through the spring and summer (depending on the species), the best season for these hollies is winter, when the colored fruit is a bright glow among the browns and grays of the branches.

There are many different deciduous hollies that vary in ultimate size and shape, exact fruit color, fruit retention, foliage quality, and ease of propagation. All of the deciduous hollies are hardy to at least zone 5, with one, _Ilex verticillata_, hardy through zone 3. As with all hollies, male and female flowers are borne on separate plants. Therefore, for good fruit set, a male plant must be near the female plant and must flower at about the same time in order for the female plant to set large numbers of fruit. The wide range of flowering times among hollies means it is important to have a male plant that flowers when the female plants do.

Winterberry, _Ilex verticillata_, is one of the hardiest (to zone 3) and most widely grown species of deciduous holly. This species is native to the eastern half of North America and can be found growing in wet, swampy areas from the Great Lakes, St. Lawrence Seaway, and Nova Scotia to the northern edge of Florida. It is a small, oval to rounded tree reaching 8–15 feet (2.4–4.5 m) in height with a spread of about 10 feet (3.1 m). Winterberry is relatively slow-growing and tends to form multistemmed clumps with dark, charcoal to medium gray bark. The foliage is an attractive deep green that does not develop fall color. Early in the fall, winterberry's small berries ripen into dense, colorful clusters, which are brilliant against the foliage until leaf-fall and then contrast effectively against the dark stems.

Winterberry should be transplanted balled-and-burlapped. It prefers moist, acidic soil conditions, but it will tolerate a range of soils as long as they are not exceptionally basic. Full sun will result in the best flowering and fruit set, but partial shade is also acceptable. *Ilex verticillata* is propagated by softwood cuttings, which are taken throughout the summer, treated with rooting promoters, and rooted under mist. There are many, many good cultivars of winterberry. Two excellent selections are 'Winter Red', a widely available scarlet cultivar with beautiful, dark green foliage and intensely colored fruit, and 'Winter Gold', a somewhat compact plant with pinkish gold fruit and light green leaves.

Possumhaw, *Ilex decidua*, is native to the southeastern United States. It differs from winterberry in height, branching habit, and bark color. Possumhaw is taller than winterberry, with the potential to reach 20 feet (6.5 m), and it has more intricate branching. The bark is a beautiful, very light gray. *Ilex decidua* tends to be less hardy than winterberry (to zone 5), and it is more tolerant of basic soils. Possumhaw is more difficult to propagate vegetatively than winterberry; it is often grafted. The foliage is a handsome dark green, and turns a light yellow in the fall. It generally flowers at the same time as American holly, *Ilex opaca*, and so usually is successfully pollinated by *I. opaca* males. As with winterberry, there are great numbers of good cultivars. 'Byer's Golden' is an unusual yellow-fruited cultivar with bright yellow fruit. 'Warren Red' fruits heavily with bright red fruit and may be somewhat more hardy than other possumhaw selections.

Probably the best available deciduous hollies are the hybrid cultivars of *Ilex verticillata* and the Japanese native finetooth holly, *I. serrata*. Hardy to at least zone 5, these hybrids exhibit prolific fruiting, excellent fruit color and color retention, and have handsome growth habits. Fruit colors early in the fall, so that the berries are effective before the leaves fall, making a lovely combination with the green foliage. 'Sparkleberry', from the United States National Arboretum, has an upright habit and very persistent scarlet fruit. 'Bonfire' (Plate 182) is a stunning selection with brilliant, true red fruit in dense clusters. 'Apollo' is a good male hybrid to plant with 'Sparkleberry' and 'Bonfire'. 'Carolina Cardinal' is a special hybrid selection from the North Carolina State University Arboretum released jointly with the North Carolina Association of Nurserymen. It is a very compact cultivar with beautiful dark brown bark that sets a striking framework for the masses of crimson berries clustered along the branches. Low-growing and full, this selection will be perfect for smaller winter landscapes.

The many cultivars of deciduous hollies offer a variety of color and character for any garden. A large massed planting near a pond or stream is an ideal way to show off the brilliant fruit to good advantage. But even a single specimen, carefully situated at a focal point of the garden, will create a beacon of color unparalleled in the winter landscape.

Ilex cornuta 'D'or' and 'O'Spring' / 'D'or' and 'O'Spring'
Chinese hollies   PLATES 183, 184

"Evergreen" is a word used over and over in association with the winter landscape. Plants that hold their foliage all through the winter have been valued since gardens began in temperate regions for the year-round beauty they add at the otherwise bare time of year. Of the evergreens, hollies (*Ilex* spp.) have been among the most valuable for their reliable, distinctive foliage and form in variations on a dark green theme, and for their bright fruit. But not all hollies offer the predictable combination of green and red. Some have variegated foliage and differently colored fruit that add a bright surprise to winter landscape. Two of the brightest, and toughest, are *Ilex cornuta* 'O'Spring' and *I. cornuta* 'D'or', two unusual cultivars of Chinese holly.

'D'or' and 'O'Spring' Chinese hollies bring a welcome splash of bright golden color to liven up the winter garden. The dark green, shiny foliage of 'D'or' is the perfect backdrop for its abundant crop of yellow-gold fruit. The berries gleam delightfully in the winter landscape, complementing other yellow elements like the branches of yellow-twig dogwood (*Cornus sericea* 'Flaviramea') and golden conifers. Branches of the golden berries make excellent accents for holiday decorations, both indoors and out. 'O'Spring' goes even further with a golden show. Its flashy foliage is variegated with cream and gold, so that it looks as if it has been tie-dyed with light yellow, cream, and bright gold, with almost no green left showing.

These two cultivars are selections of the old reliable Chinese holly, an incredibly heat-tolerant evergreen that can be relied upon to stand up to the toughest conditions, from urban wastelands to soils that should really be made into bricks. In appearance, the Chinese holly species itself is a less than elegant, coarse holly with severely dangerous, long-tipped spines on exceptionally stiff foliage (anyone who has ever planted a hedge of Chinese holly is never quite the same). Cultivars of Chinese holly have been selected to minimize the spiny character of the species and maximize other desirable traits, such as dark green foliar color and abundant, bright-colored berries.

'D'or' and 'O'Spring' Chinese hollies are a refreshing alternative to the ubiquitous 'Burfordii' and 'Burfordii Nana', which can be found, it seems, in 90% of all commercial landscapes in the American South. 'D'or' originated as a sport, a branch mutation of 'Burfordii'. It is almost identical to its parent in most respects, with dark black-green, highly glossy leaves (tipped with only one, reduced spine), a densely rounded habit, excellent heat tolerance, general landscape adaptability, and good fruitfulness. The exciting difference is that 'D'or' bears bright golden berries (Plate 183), while 'Burfordii' has the standard scarlet. New foliage emerges tinged with burgundy but soon loses that character.

'O'Spring' grows very densely in an irregularly pyramidal shape. This cultivar is one of the showiest hollies well adapted to stressful sites with heavy soils and hot summers. Color is best with some afternoon shade (to prevent any minimal scorch that might occur on the lighter areas of the foliage), but it remains in good condition all year (Plate 184). 'O'Spring' is relatively slow-growing but is not a dwarf. It will eventually reach 15 feet (4.5 m). When planted among dark evergreens, 'O'Spring' becomes a marbled work of botanical sculpture.

All hollies are dioecious; that is, the male and female flowers are borne on separate plants, and both male and female plants are required for good fruit set on the female. Females of a given species can usually be pollinated by other species whose males are in bloom and shedding pollen at the same time that the female flowers are at anthesis. So as long as there are a few male hollies in bloom in the neighborhood where your fruit-bearing holly is flowering, you should get good fruit set.

In general, Chinese holly should be transplanted from a container into full sun or partial shade. Almost any soil will work, except those that are underwater for extended periods. *Ilex cornuta* is fully cold-hardy to zone 7 and sheltered areas of zone 6 but may suffer foliar damage in severe winters unless planted in a protected site. It will generally recover from brief periods of severe cold but may lose some foliage and suffer some twig kill. There are no serious pest or disease problems with these cultivars.

Chinese holly is readily propagated from hardwood cuttings taken almost any time some hardened wood is available. Cuttings are treated with rooting promoter and rooted under mist. Propagation of hollies from seed can be a challenge, and Chinese holly is no exception. Seed is separated from the flesh of the berry, sowed in a moist, well-drained medium, and kept somewhat moist. It can take 18 months for germination. It is important to remember that propagation from seed will result in plants with a range of horticultural characteristics, including whether plants are male or female. Cultivars must be propagated from cuttings to retain their unique character.

The exceptional Chinese hollies 'D'or' and 'O'Spring' bring a flash of color to the winter garden. The rich golds of their fruit or foliage are gorgeous, and the display is dependably consistent in the face of extreme heat and other demanding site conditions. At the North Carolina State University Arboretum, both 'D'or' and 'O'Spring' shine in the gardens of the east arboretum. They are refreshing alternatives to other hollies anywhere a broad-leaved evergreen is desired. It is especially effective to use 'O'Spring' and 'D'or' in small numbers mixed with other evergreens for a splash of color. The slow-growing 'O'Spring' makes an excellent evergreen for gardens with a little shade and restricted space. Its neat and tidy silhouette is manageable in small gardens, and it brightens dark corners. A walk past these hollies in winter reveals a welcome sparkle of gold.

Jasminum nudiflorum / Winter jasmine   PLATE 185

Because they are not blooming during the traditional spring peak at garden centers and nurseries, many of the plants that flower in January and February are difficult to market. One readily available plant that often is overlooked for brightening winter gardens is *Jasminum nudiflorum*, winter jasmine. From midwinter to early spring, the branches sparkle with lemon-yellow, forsythia-like blooms with a light, elusive fragrance.

This charming relative of the olive tree is native to China. It is a spreading fountain of arching branches that creates a rounded, many-stemmed shrub about 4 feet (1.2 m) high to 6 or 7 feet (1.8–2.1 m) wide. The fine, willowy stems are a bright grass-green, a refreshing character in itself in winter. Winter jasmine's flowers are a more pastel yellow than forsythia and open more gradually throughout the blooming period, so they have less of a mass effect than forsythia. The deciduous foliage of winter jasmine is also quite lovely. The glossy, emerald-green leaves are divided into three leaflets. The foliage does not develop any fall color.

Jasminum nudiflorum is well adapted to a wide range of conditions, from sun to shade (although flowering is reduced in shade), and from rich, loamy soils to poor, dry ones. It will perform well in heavy clays. Winter jasmine is completely hardy to zone 6, with survival but poor flowering and significant stem kill in zone 5. The plant is a fast grower that spreads by rooting in at the ends of its branches. The ready growth of the shrub can be an advantage in difficult sites, but if the shrub becomes somewhat invasive, it is not difficult keep ahead of with regular cutting back or root pruning. Periodic severe pruning, cutting branches to within 1 foot of the soil just before new growth begins in the spring (similar to pruning *Buddleia* spp.) will help keep the plant vigorous and well-shaped. Winter jasmine is very easy to propagate from cuttings taken in summer and rooted under mist. There are two rare cultivars of winter jasmine, 'Aureum', with gold-variegated foliage, and 'Variegata' with white-margined leaves.

Winter blooms are always an unexpected pleasure in the garden. *Jasminum nudiflorum* parades its pastel yellow blooms in cheerful cascades, even in the most demanding of sites. It is an excellent choice for banks and masses on tough slopes and is spectacular cascading over a wall. At the North Carolina State University Arboretum, winter jasmine brightens the paths in several areas of the east arboretum, reminding us of what a pleasure the winter garden can be.

Juniperus conferta 'Silver Mist' / 'Silver Mist' shore juniper   PLATE 186

Shore juniper is a well-known evergreen groundcover in the United States, particularly in coastal regions. Native to the coastal areas of Japan, it shows tough persistence in the face of salt spray, dry conditions, and low fertility. It is a popular choice for stressful sites. A number of attractive cultivars of the species are available, but 'Silver Mist' stands out from all the rest. This unique cultivar displays silvery foliage and an especially dense growth habit that does indeed create the illusion of a silver mist over the ground.

Like all shore junipers, 'Silver Mist' is a low-growing, broad-spreading, evergreen conifer. Unlike other shore junipers, the needles of 'Silver Mist' are covered with a very heavy waxy bloom, which gives the plant its silver-gray color. In winter, the silver is augmented by a coppery plum as some of the foliage changes color with the colder weather.

'Silver Mist' grows in an extremely dense, mounded shape that effectively covers the ground. It will eventually grow to a height of approximately 1 foot (0.3 m) with a 3–4 feet (0.9–1.2 m) spread (depending on site conditions). 'Silver Mist' is more compact and prostrate than other shore juniper cultivars and is somewhat slow-growing, but the quality of its foliage is exceptional. Many shore junipers are subject to foliar die-back, particularly in poorly drained areas, but 'Silver Mist' shows little die-back and is more tolerant of heavy clay soils than many other cultivars.

Shore junipers may be somewhat more shade-tolerant than other junipers, but full sun is best for 'Silver Mist'. Like all junipers, 'Silver Mist' performs best in a site with excellent drainage, but it also performs well in clay soils with only moderate drainage. As a shore-adapted plant, this juniper will continue to prosper in windy, dry conditions and is very tolerant of salt spray. However, this does not mean that coastal regions are the only areas to grow 'Silver Mist'. It is generally hardy to zone 6, but temperatures below −10°F (−21°C) will likely cause some foliar damage. Damaged foliage and twigs can be pruned off, and the plant will recover. Shore juniper cultivars, including 'Silver Mist', are propagated by cuttings. Cuttings with a little wood on them are taken in January and February, after several killing frosts. The cuttings are treated with rooting promoters and rooted under light mist.

'Silver Mist' shore juniper creates a beautiful mound of silvery gray in the garden. Its adaptability, compact habit, and dense, quality foliage make it a tough yet well-mannered groundcover. It makes an excellent massed planting on a demanding bank or slope or in rolling areas of the garden. It is attractive in small grouped plantings as a foundation for larger shrubs and trees. 'Silver Mist' can also be a wonderful, contrasting groundcover under golden or dark green conifers, and it is lovely topping the edges of raised beds. At the North Carolina State University Arboretum, 'Silver Mist' adds a luminous glow to the garden's grounds in the mixed shrub border, and in the conifer collections. This fine cultivar holds many possibilities for the landscape, especially in the quiet season, when its subtle color is greatly appreciated.

Juniperus virginiana / Eastern red cedar   PLATE 187

The distinctive upright form of our native eastern red cedar, *Juniperus virginiana*, punctuates the fields and roadsides across the eastern half of the United States. There's something kind of special about an old field or hillside dotted with eastern red cedars. Their conical shapes have a mischievous, elfin feel about them.

Eastern red cedar carries something of an undeserved reputation for being weedy—probably because it is so incredibly tough and dependable. Generally seen as a 20–40-foot (6.2–12.4 m) evergreen tree (although it can grow as tall as 100 feet [31 m]), *Juniperus virginiana* has an upright, pyramidal shape when young, and becomes more irregular and slightly pendulous as it ages. The lower limbs die off as the tree matures, revealing its beautiful sinewy trunk, with silver-gray bark that peels in long strips to reveal a reddish brown inner layer. Adult needles are very fine textured and are pressed flat against the stem. Juvenile foliage is short and prickly and bluer in color than adult needles. The new growth of the foliage is emerald-green in spring, aging to silver-gray in summer and darkening in the winter, sometimes almost to a plum-black.

Eastern red cedar is a dioecious species; i.e., the male and female cones are on separate plants and both are required for fruiting. The female cones resemble tiny berries and have a waxy, bluish bloom that is incredibly beautiful. A heavily fruiting eastern red cedar appears veiled with its own entrancing blue glow. Eastern red cedar tends to bear heavily in alternate years, so the fruit on female trees is very conspicuous every other year.

Eastern red cedar is extremely hardy (to zone 2). It will tolerate many difficult sites, including poor, gravelly soils; acid or basic soils; and drought, heat, and severe cold. It prefers an open, airy, sunny site and does not do well in the shade past the seedling stage. There are two problems to note: bagworms and cedar apple rust. *Juniperus virginiana* is moderately susceptible to bagworms, but the bags can be removed by hand or appropriate pesticides may be used.

Cedar apple rust is a serious problem in apples and hawthorns, and eastern red cedar plays a role in the spread of the disease by serving as an intermediate host. The galls that form on cedar trees cause no damage to the cedar tree itself, and they can be quite interesting, especially when their long, feathery, bright orange fruiting arms appear. But because of the severe effects of the disease on apples and hawthorns, it is not advisable to plant eastern red cedar in the vicinity of apples or hawthorns. However, this does not mean that eastern red cedar should not be planted at all. In regions with no commercial orchards, or where orchards are routinely sprayed because of wild populations of eastern red cedar, it is safe to plant this excellent tree. The spores of cedar apple rust can travel as far as 2 miles on the wind, so if you garden near commercial nurseries, make sure to determine if there are already plantings of eastern red cedar in your vicinity before planting.

Many cultivars of eastern red cedar are available, some low and spreading, some very upright. Some cultivars offer exceptionally blue foliage (similar to eastern red cedar's western cousin, Rocky mountain juniper, *Juniperus scopularum*). *Juniperus virginiana* 'Grey Owl' (Plate 187) is an unusual cultivar that grows with a wide-spreading, bushy habit reminiscent of the well-known Pfitzer junipers. Its foliage is a silvery blue-gray that contrasts marvelously with the russet hues of many deciduous plants in winter. It can ultimately reach 6–8 feet (1.8–2.4 m) in height with an equal or greater spread, but is usually seen at heights near 5 feet (2.5 m). 'Grey Owl' may actually be a hybrid of the eastern red cedar cultivar 'Glauca' with a Pfitzer-type (*J. chinensis* 'Pfitzeriana') selection. Most of these cultivars are difficult to find in the nursery trade, but they can be mail-ordered from specialty nurseries. Native seedlings can also be propagated, without excessive difficulty, from seed or vegetatively. The bluish "berries" can be collected from wild trees in winter, layered in moist peat at 40°F (4°C) for 3 to 4 months, and sown. Seedlings will be quite variable. Vegetative propagation by grafting or layering will insure retention of desirable characteristics. If you choose to propagate a wild tree, make sure to observe it in all seasons of the year. Some cedars are prone to bronzing of the foliage in winter, which can be unattractive, and it's best to select a dark green tree that does not exhibit this trait.

The humble eastern red cedar has been a traditional indoor Christmas tree since the custom began in the 1800s. All winter long, *Juniperus virginiana* holds its appealing blue fruit, providing food for the birds and delighting our eye as much as any decorated Christmas tree.

In the outdoor garden, the dark winter foliage and strong form of the tree are dramatic. Eastern red cedar is an excellent tree for naturalizing in old fields or meadows, especially on poor soils or near water. There is something about the color and shape of this tree that conveys a sense of spirit and personality unlike any other. One of the most stunning sights I know is a combination of eastern red cedar with the winter-red tones and texture of broomsedge, reflected in a nearby pond or bay.

Mahonia spp. / Mahonias   PLATES 188, 189

While accidentally eavesdropping on a tour group at the North Carolina State University Arboretum, I heard one of the most wonderful descriptions of *Mahonia*. Upon coming around the corner of the shade house where much of the *Mahonia* collection resides, a delighted visitor gasped, "Oh, a fountain of gold!"

Indeed, *Mahonia* in flower is a sparkling fountain of bright gold for gardens in winter. Multiple upright, elongated clusters of hundreds of small, bell-shaped, yellow flowers burst up from the growing points of the shrub and arch over in a shower of gold with a light, sweet scent. The leaflets of these evergreen shrubs distinctly resemble holly leaves, but that is where the resemblance to holly ends. *Mahonia* does develop showy, berrylike fruit, but the purple-black berries are olive-shaped and borne in clusters like bunches of grapes.

Most mahonias are multistemmed shrubs reaching 6 to 8 feet (1.8 to 2.4 m) in height with a spread of 5–6 feet (1.5–1.8 m). The plants have a unique many-layered, relatively coarse habit. The compound foliage consists of a 1–2-foot-long (0.3–0.6 m) main stem lined with two ranks of dark green leaflets that are very similar to the leaves of hollies (hence, the common name "Oregon grape holly" for *Mahonia aquifolium*). Cold-hardiness varies with the species; some of the most refined and beautiful forms are also the most tender. The yellow flowers of *Mahonia* are its star attraction. The time of flowering and degree of fragrance and showiness also vary with the species and cultivar, but all mahonias are lovely in flower.

Mahonia does best in shady sites on loamy, acid, well-drained soils, but it will tolerate heavy clay as well. The plant does best in some shade. Full sun in winter will result in bronzed and burned foliage. Too much lime can cause chlorosis.

Mahonias can be propagated from seed, which requires 3 months of cold treatment layered in a moist medium. Named selections are propagated from cuttings taken in early winter, following some cold weather, and rooted under mist. A cutting consists of an entire leaf, with all of its leaflets and including the main stem joint (the primary leaf node) at the base of the cutting. Older plants can be successfully divided as well.

At the North Carolina State University Arboretum in Raleigh, North Carolina, J. C. Raulston has grown and evaluated many different mahonias for their hardiness, flowering and foliar characteristics, and performance. Many *Mahonia* species, hybrids, and cultivars are fine candidates for the winter garden.

One of the most readily available is *Mahonia bealei*, leatherleaf mahonia (Plate 188). This species flowers profusely in January and February in zone 7. The flowers are quite fragrant, especially when there are several plants together. As the fruits mature, they develop a blue, waxy bloom over their black skin, which gives them a lovely soft color. The texture of leatherleaf mahonia is coarse but interesting. As the plant gets some size to it, the foliage acquires an architectural quality that will draw the eye away from other plants with more subtle charms, something to keep in mind when selecting a site. Leatherleaf mahonia will usually reach 3–6 feet (0.9–1.8 m) in height but can get up to 8–10 feet (2.4–3.1 m). A native of China, it was introduced to America in the mid-1800s. It is hardy to zone 7; to zone 6b, in sheltered sites.

Mahonia aquifolium, Oregon grape holly, is wonderful in many parts of the United States but is not the best choice for areas with hot, wet summers, where it languishes. Native to British Columbia and the northwestern United States, Oregon grape holly has very glossy, blue-green foliage. Many cultivars are ascribed to this species, but there is a good deal of confusion over the cultivar origin of many of these. Oregon grape holly is the most cold-hardy *Mahonia* (to zone 4), tolerating temperatures to −15°F (−23.5°C), but there is some variation in cold-hardiness among the cultivars of this species.

Mahonia fortunei, Chinese mahonia, has especially attractive foliage and is relatively compact in habit. It is one of the least cold-hardy of the mahonias, being reliable only to the warmer areas of zone 8; temperatures in the neighborhood of 0°F (−16°C) will cause damage.

Mahonia japonica, Japanese mahonia, is very similar to leatherleaf mahonia but is somewhat more cold-hardy (to zone 6).

Mahonia lomariifolia is a very tender species (to zone 9) with perhaps the most beautiful foliage of the genus. The very long leaves, with narrow, fine-textured, glossy emerald-green leaflets, resemble some type of giant fern. This species has been crossed with both *M. bealei* and *M. japonica* to create a number of excellent, more cold-hardy hybrids. 'Arthur Menzies', a hybrid of *M. lomariifolia* and *M. bealei*, is probably the most floriferous of the mahonias. It blooms around the winter holidays with dense, strongly upright clusters of vivid canary-yellow flowers (Plate 189). It is a show-stopper for the shady areas of the winter garden.

Mahonia × *media* are a series of hybrid cultivars created in England with crosses between *M. japonica* and *M. lomariifolia*. These hybrids are notable for their midwinter flowering with large masses of fragrant yellow flowers. Any of these selections are excellent plants for gardens in zones 8 and warmer. 'Buckland' has huge flower clusters of a dozen upright stems. 'Charity' has softly arching clusters of clear canary-yellow. The beautiful foliage of 'Faith' resembles that of its *M. lomariifolia* parent. 'Lionel Fortescue' is another selection with very showy flower clusters. 'Winter Sun' has particularly upright, stiff flower clusters.

Mahonia nervosa, Cascades mahonia, is lower and denser in habit than other mahonias. Its more graceful, elongated foliage is excellent in combination with other mahonias. Cascades mahonia is native to the western American continent and relatively cold-hardy, being reliable in zone 5.

Mahonia repens, creeping mahonia, is also a western American native. It is a somewhat upright groundcover that does exactly that and does not bloom until early spring. It is also hardy to zone 5.

The many *Mahonia* selections offer bright, primary color combined with bold form and texture for winter gardens in the milder parts of the United States. They are some of the best plants for small, shady urban gardens, both walled and hedged, and they are a fabulous way to add bold texture to evergreen mass, by interplanting with other broad-leaved evergreens. Their yellow-gold flowers are a natural with yellow-variegated conifers and hollies. Mahonia is an excellent companion to perennials in a border or bed as it extends the bloom season and adds year-round interest. The unique foliage, topped with great splashes of yellow, creates an unusual effect among the subdued forms and hues of other deciduous plants in the garden.

Morus australis 'Unryo' / Contorted mulberry   PLATE 190

The winter-bare branches of *Morus australis* 'Unryo', contorted mulberry, create a twisting, turning sculpture of nature that is a spectacular addition to the garden in winter. Previously classified as *Morus bombycis* 'Unryo' and sometimes referred to as the cultivar 'Tortuosa', this native of China is a relative of the more common red and white mulberries, which bear tasty fruit in the summer. Contorted mulberry resembles its cousins only in the summer, when its twisted branches are covered with very large, dark green leaves. This plant's best season is winter, when the long, thick, corkscrewed branches can be seen clearly.

Contorted mulberry is a shrubby, multistemmed small tree, cold-hardy to zone 6. It grows into a rounded mass of branches, eventually reaching 20–30 feet (6.2–9 m) in height with a broad spread of 15–20 feet (4.5–6.2 m). A rapid grower, it can add as much as 8–10 feet (2.4–3.1 m) in one year. The leaves may be 6–8 inches (15.2–20.3 cm) long, and they may develop bright gold fall color in some years but often the leaves drop before good fall color develops. Contorted mulberry prefers full sun or light shade and will do best in moist, well-drained conditions but is a wonderful plant for winter interest on demanding sites where other plants would die. This is a very tough plant that thrives in a variety of soils, from heavy clay to dry soils, and in salty, seaside environments.

Morus australis 'Unryo' is readily propagated from cuttings taken any time of year and rooted under mist. Because it is so readily propagated, it is often more reasonably priced than many other desirable contorted plants that must be grafted, such as *Corylus avellana* 'Contorta', the well-known Harry Lauder's walking stick. The branches of contorted mulberry are not as intricately twisted as that plant and are somewhat more coarse, but what contorted mulberry lacks in refinement it makes up for in vigor and affordability.

Contorted mulberry's rapid growth rate, adaptability to a wide range of landscape conditions, and fascinating winter character even as a young plant make it an excellent choice for new gardens in recently developed residential areas, which often have less than ideal sites. Its wild branching combines well with many conifers and makes an interesting frame for early spring-flowering bulbs, such as crocus and early daffodils, which bloom before contorted mulberry's leaves are out. The interesting, adaptable contorted mulberry is a plant that can bring a new dimension to winter in any garden.

Nandina domestica / Heavenly bamboo, Nandina   PLATES 191, 192

Nandina domestica is the embodiment of quiet grace in the garden. Although it is actually not a true bamboo at all, the outline of nandina's foliage and the shape of its woody stems are reminiscent of many forms of bamboo. But it is in the late-season garden that heavenly bamboo really shines. In fall, heavy bunches of bright scarlet berries are draped like miniature crimson grapes across the vivid foliage, and the bright fruits persist well into winter. Among the most striking of displays for fall and winter gardens, nandina rivals even the deciduous hollies for their show.

Heavenly bamboo is an evergreen, shrubby relative of the barberries (*Berberis* spp.) that develops into an upright, unbranched shrub of 3–9 feet (0.9–2.7 m) in height (depending on cultivar). *Nandina* brings along none of the difficulties of culture or invasiveness problems that often accompany many of the nicest true bamboos. Over time, heavenly bamboo spreads relatively slowly to create colonies by suckering, but the beautiful, soft masses of plants are easily controlled.

The leaflets of the divided leaves are shaped in the narrow, pointed shape of typical bamboo leaves. New foliage emerges tinted wine-red, then turns a deep emerald- to olive-green as it matures, and finally takes on burgundy hues again in the fall with the return of cool weather. The burnished foliage sweeps down from the plant's apex in a lovely mosaic of glossy olive green and burgundy-tinted leaflets. In late spring and early summer, delicately branched clusters of pink-tinged flower buds wave from the branches. Each bud will open into a lacy white bloom, 0.5 inch (1.2 cm) across. The flowers mature into eyecatching clusters of bright red fruit (or, in the case of one form, ivory-colored fruit).

One of the most useful characters of heavenly bamboo is its ability to thrive under almost any landscape conditions. While it prefers moist, fertile, well-drained soil, *Nandina domestica* will also perform well in heavy, poorly drained clays and light, drier soils. It will flower and fruit reliably in both full sun and heavy shade (although flowering and fruiting will be the most prolific with good sun) and even performs well underneath large shade trees. Heavenly bamboo is reliably hardy to the warmer parts of zone 6. In the colder areas of zone 6, severe weather may result in foliar damage but should not kill back the entire plant.

Cultivars are propagated from hardwood cuttings and from tissue culture, but the species, and its ivory-fruited form, can be propagated from seed. Seed is cleaned from the soft berries before sowing in the late fall or winter, and it may take as long as a year to germinate. Foliage color, abundance of fruit, and overall habit can be very variable among seedling plants. Because of this, it is unusual to find the species *Nandina domestica* in nursery production and available in the trade, although wonderful old species plants can still be seen in older gardens. The species can create some of the loveliest plantings of heavenly bamboo to be found in the landscape, but the numerous excellent cultivars available are all consistent performers in terms of their own unique landscape traits. It is important when looking for heavenly bamboo cultivars to distinguish between the low-growing dwarf types and the more traditional upright forms that will eventually create a mass.

Many interesting full-sized forms are available. 'Alba' has cream-colored berries, and its green foliage lacks the red tones of other forms. This form comes true from seed, unlike other named varieties. 'Moyers Red', which will eventually reach 5–6 feet (1.5–1.8 m), is slower growing than many seedlings. It has very pink flower buds and intense crimson fruit. 'Royal Princess' has a very upright, narrow habit to 6 feet, and smaller leaflets than the species. 'Umpqua Warrior', one of the largest, reaches to 8 feet (2.4 m) in height with exceptionally large flowers and fruit.

There are numerous dwarf and medium-sized forms with compact habit and various color traits that make excellent landscape plants in small groups and masses. 'Fire Power' (Plate 191) has a very compact habit to 2 feet (0.6 m) in height and width. The very dense foliage turns a neon red in fall and winter, and retains much of that fire-engine character even in the summer. 'Gulf Stream' reaches slightly larger size than 'Fire Power' but is slightly less electric-red than that cultivar in the winter and turns more green in the summer. 'Harbour Dwarf', which forms a tight mound to 2 feet (0.6 m), has dwarfed flower clusters and similar leaf coloring to the species. It is not a show-stopper like 'Fire Power', but it is perhaps the most graceful of the dwarf forms. 'Okame' is more upright in shape, reaching 3 feet (0.9 m) tall, and has good, bright red winter color. 'San Gabriel' (Plate 192) offers very compact, low habit combined with the most unique foliage of any readily available form. Its leaflets are so narrowed that they look almost like grass blades, giving a very fine texture in the landscape. The foliage acquires good burgundy-red winter color. 'Umpqua Chief' reaches medium height of 3 to 5 feet (0.9 to 1.5 m); 'Umpqua Princess' matures to only 3 feet (0.9 m). 'Wood's Dwarf' is possibly the most compact form of all those in the trade; it has bright red winter foliage.

A number of rare dwarf Asian cultivars of heavenly bamboo have leaves that consist only of the central tissue around the midrib of the leaflets, never developing any blades. These incredible plants look like small and elegant sculptures made of many fine olive-colored branches. Their limited amount of leaf tissue results in very slow growth; 40-year-old plants may be as little as 25 inches (63.5 cm) tall! Some of these rare forms can be viewed in the shade house at the North Carolina State University Arboretum, while many of the other named cultivars line the Arboretum paths leading out of the white garden through the cypress and witch hazel collections of the east arboretum.

Heavenly bamboo meets the challenges of less than ideal conditions with the refinement and grace of the great bamboos. Large nandina plants make eloquently graceful screens and masses, where they are also a ready source of holiday greens. Nandina's Asian character works well in Japanese-style gardens. The muted tones and fine texture are easy to mix with many different plants, from grass to perennials to small trees and shrubs of all types. They are especially appealing in winter when combined with grass. The warm-toned brown and beige persistent blades of grass create artful silhouettes against the dark green, fine-textured nandina foliage, and the composition is punctuated by nandina's bright fruit.

Phellodendron amurense / Amur corktree   PLATE 193

The transitional season from fall into winter brings out the somber side of the landscape, when the newly bared branches become the perfect frames for any remaining foliage or fruit. The persistent clusters of shriveled black fruit that cover the crown of *Phellodendron amurense*, Amur corktree, in late fall and winter are an interesting feature in the late year's garden, once the distracting leaves are blown away by the breeze. The furrowed, light gray bark of this tree is another handsome feature for the winter garden, especially on old trees, which develop dramatically deep ridges and furrows in corky patterns. The trunk of Amur corktree is substantial compared to the tree's relative height, so the bark is an important ornamental feature in the winter.

Phellodendron amurense is a large, deciduous tree from China that will reach 25–40 feet (7.5–12.4 m) in height with an equal spread. The slow-growing crown is open and somewhat spreading, with an attractive branching pattern. The leaves are divided in a fashion similar to those of walnut or pecan. Foliage is a dark, shiny green throughout the summer and turns a handsome deep yellow in the fall. Amur corktree's yellowish flowers are borne in inconspicuous branched clusters in late spring and are not especially showy. However, on female trees, the flowers mature into small, round, black, berrylike fruits about 0.5–1 inch (1.2–2.5 cm) wide. As the fruits hang on the branches, they dry and shrivel a bit until they begin to resemble large, plump raisins.

Amur corktree is a relatively tough tree that will tolerate a range of soils from clays to light sand. It transplants well into most sites and does best in full sun. There are no significant pest or disease problems. The tree is completely hardy to zone 3. It is readily propagated from direct-sown seed, which germinates easily and quickly.

'Macho' (a non-fruiting male, of course) is the only cultivar of *Phellodendron amurense*. Its crown is more spreading than that of the species, and its bark is especially corky. There are also two other rare species that you may encounter in botanic gardens and arboreta. *Phellodendron lavallei*, Lavalle corktree, is native to Japan and has more upright branching and less corky bark character than *P. amurense*. *Phellodendron sachalinense*, Sakhalin corktree, is native to Korea and parts of Japan and western China. It is more upright and arching in habit and larger than its botanical cousins.

A view of the winter sunset through the branches of Amur corktree reveals an unusual tracing of line and form (Plate 193). This whole sculpture of fruit and branch is hung on a massive trunk with its own appeal. As a specimen to add interest to the winter garden, *Phellodendron* is unmatched by most other large deciduous trees. It is a unique lawn tree with exceptional textural and architectural appeal year-round. *Phellodendron amurense* is one of the most striking of the many textures and forms that make up the quiet magic of the plant world, now prepared for winter's garden.

Picea omorika / Serbian spruce   PLATE 194

Conifers are the backbones of many modern landscapes. Because gardeners rely so heavily on this large group of plants, there is a great need for an equally large selection of coniferous material to choose from. This is especially so when considering the diversity of site conditions, area, microclimates, and gardeners' tastes. One out-of-the-ordinary conifer that offers an alternative appeal for gardens both large and small is *Picea omorika*, Serbian spruce.

Serbian spruce, as the name suggests, is native to a small area in the general region of what was formally Yugoslavia, where it is found on chalky soils in mountainous areas. The native species is a tall, narrowly columnar, needled evergreen tree that can reach heights of 100 feet (31 m). It holds its branches in a remarkable way, so that the tree looks as if it were raising arboreal arms toward the sky. The characteristic habit of this tree, with its many "raised arms," is a beautiful and unique shape for large-scale landscapes, but the species is sufficiently slow-growing to remain at a scale useful for smaller landscapes for many years. The individual needles of Serbian spruce are fine-pointed and narrow, but they spread out away from the branch, giving the tree a very fine texture. Needles are generally a handsome blue-green that remains uniformly attractive through winter and summer.

In addition to the species, there are a few dwarf and slow-growing cultivars of Serbian spruce that are excellent plants for smaller gardens. While these smaller forms do not develop the unique branching character of the full-sized species, they do show the same refined texture and foliar color, and they offer their own characteristic forms that are equally appealing. 'Expansa' is a spreading dwarf reaching about 3 feet (0.9 m) in height with no true main stem or leader, so it spreads irregularly as if the lower branches of a Christmas tree were growing out of the ground. When 'Expansa' is propagated by grafting, however, it will eventually develop a leader and then grows into a compact upright selection reaching 10–15 feet (3.1–4.5 m). Both "versions" of 'Expansa' may turn up in nurseries and garden centers so make sure to determine if the plant was grafted or not. 'Nana' is a rounded dwarf reaching 8–10 feet (2.4–3.1 m) with an ultimately conical shape. Its needles are pressed more flatly to the stem than other forms.

'Nana' is an interesting alternative to dwarf Alberta spruce (*Picea glauca* 'Conica'), although it is somewhat slower in growth. 'Microphylla' is another dwarf form, with slightly shorter needles and more pendulous habit than 'Nana.' 'Pendula' (Plate 194) will eventually reach full size, but it is somewhat slower-growing than the species. It has weeping branches and sometimes weeping leader growth, making it an exceptionally graceful and lovely conifer for all gardens.

Serbian spruce has a reputation for requiring deep, moist, well-drained, loamy soils. In reality, however, it will perform well in a range of soils and conditions, from heavy clays to light sands. This spruce grows well in full sun or light, high shade. It is cold-hardy to zone 4 and can be propagated from seed or from cuttings taken in winter and rooted under mist.

Serbian spruce is one of the most stunning conifers in the landscape. *Picea omorika* seedlings can be used beautifully as a screen on a small suburban lot. In 15 to 20 years, these trees will need thinning, but they make a fine planting for the foreseeable future of that landscape. As a young tree in mass or as a screen, Serbian spruce has a relaxed outline that makes it easy to include in many styles of garden. But as it matures, this outstanding conifer displays a narrow, spire-like silhouette that makes a dramatic specimen. A mature tree with its limbs gracefully raised heavenward is a vision of strength and beauty, especially in the winter garden.

Picea pungens 'Foxtail' / 'Foxtail' blue spruce   PLATE 195

Colorado blue spruce is a striking addition to larger, more formal landscapes, especially when used in a grouping. Unfortunately, most *Picea pungens* cultivars do not perform well in areas with hot summers. Colorado blue spruce is generally not well adapted to hot nights and heavy soils. When grown in these areas, there is little new growth and the foliage does not display good blue color. 'Foxtail', however, is a Colorado blue spruce cultivar that is the exception to this rule. This heat-tolerant cultivar gets its name from the shape of the branches: The outline of the most recent year's growth resembles the brush of a fox's tail.

'Foxtail' is very heat resistant and grows rapidly, eventually becoming a large, pyramidal tree. The showy foliage is a consistent, intense blue-gray. This cultivar originated at Iseli Nursery in Boring, Oregon, where it was selected initially for its unique habit of the foxtail outline of the branches. In addition to the unusual branch outline, 'Foxtail' has a rare tolerance for high temperatures, thriving under conditions that would be inhospitable to any other cultivar of Colorado blue spruce. The cultivar has shown both vigorous growth and excellent blue foliage even in the hottest of summers. It is hardy to zone 2 and deserves attention in cold climates as well as hot areas for its reliable color and the unique silhouette of its branches. Like most blue spruce cultivars, 'Foxtail' has few pest problems and requires full sun and good drainage for best growth and foliage color. *Picea pungens* cultivars are usually propagated by grafting, but some are moderately successful from cuttings, which are taken in winter, treated with rooting promoter, and rooted under mist. The species can also be successfully propagated from freshly sown seed.

Colorado blue spruce can be difficult to use effectively because of its dominant visual impact. Yet, when placed appropriately, the vibrant blue-gray foliage can make a landscape come alive. It makes a dramatic specimen in winter gardens or is excellent combined with other blue-foliaged conifers of more informal and relaxed habit and texture. The combination of blue spruce with sculpture or other garden ornament makes an especially eye-catching winter feature. This kind of design statement has previously been difficult to achieve in hot climates, where blue spruces expire slowly after several years in the landscape. With the availability of 'Foxtail', gardeners in almost any climate can enjoy the dramatic beauty of blue spruce in the garden year after year.

Pinus bungeana / Lacebark pine   PLATE 196

The surreal mother-of-pearl bark of lacebark pine is an arboreal magnificence. On young trees, the bark is an opalescent mosaic of ivories, creams, russets, olives, and silvers, melded in patches in the manner of sycamores (*Platanus* spp.) but with an entirely different dimension of grace and beauty. From a distance, the bark gleams like mother-of-pearl, or huge opals, changing hues with every viewing angle. As lacebark pine matures, the mosaic quality gradually fades away into a startling, uniformly smooth, chalky-white bark as the tree achieves great age.

Unseen in European and North American gardens before the mid-1800s, lacebark pine has been planted for centuries in temple gardens in China, Korea, and other Asian countries. Like many other plants native to China, *Pinus bungeana* (named for its European discoverer, Aleksandr Bunge) has become an important specimen in European and North American horticulture.

Lacebark pine is a large tree, reaching upwards of 80 feet (24.8 m) with great age, but in modern landscapes it is generally seen reaching 35–45 feet (10.9–13.9 m) in height with a 20-foot (6.2-m) spread. It is relatively variable in habit as a young tree and is frequently seen with multiple trunks, which can make a handsome and interesting plant in many gardens. If pruned to a single trunk, it will grow with a formal, pyramidal outline. The dark-green, stiff-needled foliage is a striking contrast to the bark of both young and mature trees.

Pinus bungeana is very slow-growing, but it is tolerant of a relatively broad range of conditions. It prefers well-drained, loamy soils in full sun but will tolerate light sands and heavy clays. It is completely hardy to zone 4 and is quite tolerant of urban conditions. The United States National Arboretum in Washington, D.C., is home to a number of spectacular lacebark pines. As with all pines, propagation is by seed or grafting onto rootstock of other closely related pines. The seed has no pre-treatment requirement and can be sown fresh.

In either single or multitrunked form, and even as a youth, lacebark pine brings great depth of character to a garden. Its elegance of form and fascinating bark make this pine stand out in a landscape like royalty from a mob. Imagine a row of lacebark pines fronting a black granite wall, their opal columns glowing against the dark face of the wall. Most specimen trees are planted with an eye to the future, but lacebark pine rewards us with incredible, immediate beauty as well as long-lived distinctive character. It is both regal and practical for large-scale modern landscapes.

Pinus densiflora 'Umbraculifera' / Tanyosho pine   PLATE 197

Although creating an authentic Japanese garden can be a daunting task, many of the wonderful concepts of Japanese gardens can be incorporated into modern American garden design and development. Creation of contemplative spaces, after the Japanese ideal, depends not only on sound design concepts but on the plant material used (or not used!) in the garden. One exceptional plant that is a classic in Japanese-style gardens is the Tanyosho pine, _Pinus densiflora_ 'Umbraculifera'.

Tanyosho pine is a venerable, well-known cultivar of the Japanese red pine, native to Japan and Korea. The native Japanese red pine, a rather open tree with long, delicate, bright green needles, is somewhat gangly in youth but eventually grows to be quite handsome, reaching heights of 50–80 feet (15.2–24.8 m) with age. Tanyosho pine, like many cultivated selections, is quite different from its monumental parent.

The habit of Tanyosho pine is very distinctive. It grows quite slowly and remains beautifully compact, reaching heights of 10–15 feet (3.1–4.5 m) in cultivation with an equal spread, a good range for today's smaller gardens. Tanyosho pine develops a rounded crown atop picturesque branches with red-brown bark. The softly undulating canopy of this tree is the embodiment of calm beauty. Like the species, this cultivar also has very graceful, slender, rich green needles, but they are somewhat shorter and are held more tightly together on the branches, making the foliage thicker and denser while still retaining a refined image.

Tanyosho pine performs best in full sun, in well-drained, somewhat acid soils. It will grow well in other soils, including heavy but not excessively wet clays and sandy soils with adequate moisture. Pines in general, including Tanyosho pine, can be subject to a number of disease and insect problems when they are stressed. Vigorous plants grown in appropriate sites should not have significant problems. Tanyosho pine is cold-hardy to zone 3, but it will not take kindly to the extended, extreme heat of zones 9–10.

Like all clonal selections of pine, Tanyosho pine is propagated by grafting. The species, _Pinus densiflora_, can be propagated from seed, which requires no pretreatment to germinate; however, the seedling forms will be variable and will generally bear no resemblance to Tanyosho pine. Many other named forms of Japanese red pine, besides Tanyosho pine, have been selected in Japan. Two of the more frequently seen selections are 'Globosa', a short-needled, rounded, ground-hugging form, and 'Pendula', a prostrate, weeping form that flows over rocky walls or the ground in a pool of bright green.

Tanyosho pine has an ageless beauty that captures the mind and spirit when entering the garden. At the North Carolina State University Arboretum, a lovely specimen is an integral part of the plantings in the Japanese garden and has grown well in the Piedmont conditions there. Tanyosho pine is an excellent choice for small to medium-sized gardens because of its small and regular crown and mature height. The fine-textured foliage and red bark maintain quiet color and interest year-round. This is a good conifer to use as a single specimen tree to hold together shrubs and perennials. Because of its size and year-round character, it is a good choice for gardens where only a single tree can be used effectively. Even when a single tree is the focus, a quiet walk into the winter garden will allow Tanyosho pine to reveal its contemplative character.

Pinus spp. 'Oculus-draconis' / Dragon's-eye pines   PLATES 198, 199

Unusual pines are a gardener's treasure for the variety and character they add to the winter landscape, and dragon's-eye pines are just such excellent trees. The common name "dragon's-eye pine" is used for three different cultivars of three different pine species, all of which carry green needles variegated with evenly spaced, discrete bands of golden or cream (usually two wide bands). When viewed from a distance, the markings give a broadly golden-banded appearance to the whole tree, resembling the eye of a mythical dragon with two yellow rings on a green background.

All three dragon's-eye pines are dramatic landscape statements or collector's specimens. Because of their bold foliage, however, they require careful placement to avoid becoming a landscape liability in an already visually chaotic garden. They can make excellent dramatic elements for a mixed planting, particularly in late fall and winter when other, more showy plants are dull or invisible. These three pines are also excellent sources of bright color against dark evergreens, and they make unique foils for red-fruited shrubs and deciduous plants with intensely colored fall foliage.

The three dragon's-eye pines show distinct differences in habit, texture, and degree of bold color. *Pinus densiflora* 'Oculus-draconis' (Plate 198) is probably the most readily available of the three. This cultivar of Japanese red pine has fine-textured, graceful needles, 3–5 inches (7.6–12.7 cm) long, which are often slightly twisted. It is a slow-growing tree that is usually seen in the landscape at 20–40 feet (6.2–12.4 m), but it can eventually get much larger, reaching 100 feet (31 m) with great age. It has an open, almost pendulous habit when young, and as the tree matures, the trunk develops a picturesque, leaning or curving shape. The crown of the tree grows to be flat-topped and spreading, with a shape reminiscent of classical Japanese landscapes. On older trees, the bark cracks to reveal an interesting orange-colored interior. *Pinus densiflora* 'Oculus-draconis' is the most delicate in texture and foliage color of the three dragon's-eye pines, but it also tends to be the least consistent in color retention during the winter, when the light bands may turn somewhat brown. It is native to Japan and prefers a well-drained acid site in full sun. It is completely hardy to zone 4 and, like all cultivars of pine, is propagated by grafting onto species understock. At the North Carolina State University Arboretum, dragon's-eye Japanese red pine glows in the west arboretum.

Pinus thunbergiana 'Oculus-draconis' is the dragon's-eye cultivar of Japanese black pine. The 3–4-inch-long (7.6–10.1 cm) needles are relatively thick and stiffly upright and are borne on rigid, spreading branches that tend to droop eventually with age. The bright yellow bands stand out dramatically against the dark black-green of the needles. The tree is usually seen at 20–40 feet (6.2–12.4 m) in height with a broad spread, but trees can reach 80 feet (24.8 m) in height. Spread can be controlled by pruning to create a more dense specimen. *Pinus thunbergiana* 'Oculus-draconis', with its excellent, consistently bright coloration, is probably the most dramatic and difficult to use of the three dragon's-eye pines. It has the boldest texture and color and makes an exciting specimen plant, but it is very challenging to blend with other interesting garden plants. It is not as hardy as *Pinus densiflora* 'Oculus-draconis' and may be damaged by temperatures below −10°F (−21°C), which will cause needle burn but won't kill the tree to the ground. It is exceptionally tolerant of salt and dry, sandy conditions and therefore makes an excellent choice for coastal areas; it is perfect for pine aficionados gardening near the beach.

The last of the three is *Pinus wallichiana* 'Zebrina', the Himalayan dragon's-eye pine (previously classified as a cultivar of *Pinus griffithii* and sometimes called 'Oculus-draconis' as well). This is the most stately and elegant of the three, reaching 30–60 feet (9–18.6 m) in the landscape and as much as 150 feet (45.7 m) with age. The graceful, medium-textured needles arch over in cascades of green and creamy gold from wide-spreading, handsome gray branches (Plate 199). The tree matures to a broadly pyramidal form of great dignity, but even as a young tree, its habit is refined and lovely. *Pinus wallichiana* 'Zebrina' prefers well-drained, sandy soil in full sun, but it also performs well in clay. Young trees will grow as much as 1–2 feet (0.3–0.6 m) a year in a favorable site. It is helpful to transplant this tree when young to avoid serious transplant shock. It is hardy to zone 5, but it is subject to significant foliar discoloring at temperatures below −15°F (−23.5°C). Planting in a sheltered site will minimize winter browning of the needles.

All three dragon's-eye pines offer beautiful variations on the theme. Solid gold foliaged plants can be overly arresting at times, especially in small gardens. The dragon's-eye pines, whose foliage is intermittently colored, offer a less strident way to brighten winter's gardens. Their golden or creamy bands painted against evergreen foliage are dramatic against the stark backdrop of the winter landscape without overwhelming the more subtle character of the garden.

Pinus taeda 'Nana' / Dwarf loblolly pine   PLATE 200

Dwarf loblolly pine is an especially magical addition to the winter garden. Its long needles, carried in dense, brushy clusters, blend together on the low, compact, billowing crown to create an emerald cloud floating over the garden.

In many respects, *Pinus taeda* 'Nana' is much like its parent, the southeastern United States native loblolly, with relatively long, soft, grass-green needles in tufts of three, coarse reddish brown bark, and indomitable toughness in hot, humid summers and clay soils. The important difference is in growth. Common *Pinus taeda* reaches heights of 50–100 feet (15.2–31 m) in nature (attaining heights of 40–60 feet [12.4–18.6 m] in most landscapes), losing most lower branches as it ages so that its open, spreading crown is perched near the top of the plant's trunk. *Pinus taeda* 'Nana', on the other hand, reaches only 20 feet (6.2 m) after 30 years and naturally develops a much-branched, gently tiered crown that looks as is if it had been pruned diligently by several generations of Japanese gardeners. This dwarf loblolly is among the most elegant of pine selections, surpassing even the revered Tanyosho pine (*Pinus densiflora* 'Umbraculifera') for natural artistic character.

Pinus taeda 'Nana' originated from seed from a witch's broom, an area of growth on an otherwise normal plant where the branches and foliage are very dwarf and appear congested—as if one section of the plant had grown in miniature. Often, if fruit or cones develop and mature on a witch's broom, the seedlings grown from this seed will be dwarf or variant in a manner similar to the witch's broom itself. Horticulturists, including Sidney Waxman of Connecticut, have built entire careers around working with witch's brooms in conifers. A number of the dwarf conifers in cultivation today were selected from just such seedling lots.

At North Carolina State University, breeding and selecting improved strains of loblolly pine has been a part of the forestry program since the early days of the college. As part of that program, seed from a *Pinus taeda* witch's broom was grown. The resulting seedlings grew into a population of beautiful, cloud-like, very slow growing loblolly pines. Because of their slow growth rate, these dwarf trees were not deemed very useful to the foresters; however, they were spectacularly beautiful ornamentals. Fred Cochran in the Department of Horticultural Science observed what magnificent ornamentals these trees were and saved them for the future. In the early days of the North Carolina State University Arboretum, J. C. Raulston recognized their ornamental potential and kept the trees on the Arboretum grounds as an important part of the collections.

Pinus taeda 'Nana' is not only elegant but also a tough landscape plant widely adaptable to most landscape conditions of the southeastern United States, the range of its parent species. But it is also hardy through zones 6 and warmer parts of zone 5 and deserves wider use in other regions. It will thrive in soils from clay to sand and is relatively drought-tolerant once established. Unfortunately, propagation of this wonderful tree is a challenge. Since all of the existing trees are physiologically mature, they must be propagated by grafting which is slow and difficult. As a result, only a few specialty nurseries carry *Pinus taeda* 'Nana'.

Each time we enter the garden, plants like dwarf loblolly pine remind us that there is a planet full of such treasures and gifts. By planting *Pinus taeda* 'Nana', we can bring some of nature's unending variety and beauty to our own gardens and landscapes and remember just how miraculous that variety and beauty can be. This plant's evergreen canopy and form create unique structure and defined spaces—seated on a bench under the dwarf loblollies, atop a carpet of long needles, is a wonderful place to be, even in winter.

Poncirus trifoliata / Hardy orange   PLATE 201

Hardy orange is a very interesting looking tree with a dense, angular habit. The branches are bright green when young and the whole tree bristles with long, sharp, vicious thorns. Hardy orange makes a unique landscape statement, particularly in late fall when the leaves have fallen and a few yellow fruits may still persist among the deep green branches and thorns. Because of its extremely nasty thorns, hardy orange can be grown at close spacing to form an impenetrable hedge that will daunt even the most inveterate of gate-crashers. As a specimen tree, its sculptural and textural effect is very striking. Even more fascinating than the species is the contorted cultivar 'Flying Dragon', with its dramatically twisted branching and sharply curved thorns.

Poncirus trifoliata, hardy or wild orange, is a deciduous, large shrub or small tree that can reach 20 feet (6.2 m) in height and spreads from 10 to 14 feet (3.1–4.2 m). Hardy orange is a citrus relative introduced from China in the 1850s. Hardy to zone 5, it has naturalized throughout much of the southeastern United States, from Florida and Texas north and east through Virginia and into Pennsylvania.

Hardy orange blooms in the spring with small, white, fragrant flowers. The foliage is lustrous green throughout spring and summer and turns yellow in the fall. The leaves are divided into three leaflets. *Poncirus trifoliata* is a relative of true orange and bears small, orangelike fruit that ripen in September or October to a dull yellow color and are covered with a light, downy fuzz. Even when ripe, the fruits are very sour, but they make delicious marmalade and can generally be used as a substitute for lemon.

Hardy orange is tolerant of most landscape conditions. It does best in full sun in moist, well-drained, acid soils, but it will also grow in a range of difficult sites from wet to hot and dry. There are no serious disease or pest problems with this tree, and it is easy to transplant. Hardy orange can be readily propagated from seed, which is layered in a moist medium and stored at 40°F (4°C) for 90 days. This cold period is necessary for good germination. Softwood cuttings taken in summer and treated with a rooting promoter can be successfully rooted.

As winter arrives, nature's cloak of leaves and flowers is blown aside. This is the season to look in the garden for an array of line and shadow, silhouette and form. Hardy orange is one of the most intricately sculpted plants of the winter landscape. As a hedge or mass, it is an interesting and impenetrable screen. The sculptural character of its branches with their formidable thorns adds intriguing interest when used as a specimen as well. For the truly adventurous, hardy orange is fascinating combined with contorted trees and shrubs, such as *Morus australis* 'Unryo'.

Prunus mume / Japanese flowering apricot PLATES 202, 203

Japanese flowering apricot gives the kind of extravagant floral display in midwinter that most plants reserve for spring. This charming iconoclast of the garden decks its bare winter boughs with ruffled blooms of white, pink, or red that perfume the entire garden with the heady scent of sweet floral spice.

The flowers open sporadically throughout the winter and early spring in a series of flushes following cold periods that have fulfilled a "chilling requirement" the plant needs to flower. After a set of flowers open, they may, of course, be frozen by a period of severe weather, but then a new set of blooms will open into a bright wave of pastel. The single or double flowers can be white, light to dark pink, rose or red and are .75–1.5 inches (1.8–3.7 cm) in diameter. Individual flowers are similar to those of the well-known Japanese flowering cherry, *Prunus serrulata*; when they are fully open, they look like minuscule cabbage roses. At the North Carolina State University Arboretum, Japanese flowering apricot trees are planted in various places to greet garden visitors with their winter beauty and fragrance.

Japanese flowering apricot is a small tree, usually with a rounded habit (with the exception of a few cultivars). It can reach 25 feet (7.5 m) in height with a spread of about 20 feet (6.2 m). Growth is rapid on young trees but slows somewhat as the tree matures. A young tree can grow as much as 5–6 feet (1.5–1.8 m) in one year in warm climates. The recent growth on young branches can be bright green, adding to the colorful display. In the summer, the oval-shaped leaves are a light, emerald-green. The foliage does not develop significant fall color in most cases (although some leaves may turn an attractive light yellow). However, the flowers mature into small, cherrylike fruit that are a warm golden color and add ornamental character later in the season. In their natural state, these fruit are only appetizing to wildlife, as they are extremely tart.

Prunus mume is native to China and Japan. In Japan, over 300 cultivars have been developed for ornamental purposes, as well as for the fruit which is made into wine or preserved. It is reliably hardy from zones 7 to 9 and into zone 6. However, the trees have continued to prosper through temperatures as low as −7°F (−19.5°C), without significant damage to the branches. This tree is worth the risk of some unhappy moments in a hard winter for the delights of its blooms in other years.

Japanese flowering apricot will perform best in well-drained, fertile, somewhat acid soils, in full sun. It will also thrive in soils with greater amounts of clay that are less well-drained. Partial shade is acceptable, but flowering may be reduced. The tree can be pruned to maximize flowering. Cutting all the branches back after flowering the initial year, then cutting only half of the branches back each subsequent year, will ensure heavy bud set. *Prunus mume* is easily propagated from summer softwood cuttings rooted under mist. Grafting is also successful. Seed must be layered in a moist medium and stored at 40°F (4°C) for 1 to 3 months to stimulate successful germination.

Most of the Japanese cultivars of *Prunus mume* are not available in this country and, until recently, the plant itself was almost impossible to find. Now a number of specialty nurseries carry a variety of *Prunus mume* cultivars. Some of the available cultivars include 'Alba', a vigorous, single white form; 'Albo-plena', an especially early, double white form; 'Alphandii', with semidouble pink flowers; 'Benishidare', with single, small, exceptionally fragrant blooms in an intense crimson; 'Benishidori', a late-flowering, double-flowered form with dark pink buds that open to bright pink blooms fading to light pink; 'Bonita', a semidouble, rose-colored form with upright growth habit; 'Dawn', with late, very large, shell-pink flowers; 'Matsubara Red', with dark red, double flowers in midwinter and an upright habit; 'Omoi-no-mama', with late, semidouble, small white flowers; 'Peggy Clarke', a double form with dark rose flowers; 'Pendula', with single or semidouble pastel pink flowers on weeping branches; 'Rosemary Clarke', with very early, large, semidouble, very fragrant white flowers with a red calyx; 'Viridicalyx', a double white form with a green calyx; and 'W. B. Clarke', another weeping form with double pink blooms.

Japanese flowering apricot is a very special ornamental tree for the winter garden. It makes an excellent specimen in walled areas where the fragrance can linger, or is lovely planted by a walk or patio where wintertime passersby can easily see the lovely flowers. Flowering branches of *Prunus mume* also make wonderful additions to cut floral designs. One or two branches can beautifully perfume an entire room.

Sciadopitys verticillata / Japanese umbrella pine   PLATES 204, 205

Many rare and interesting conifers grown in our landscapes should remain undisturbed so they can be appreciated for their incredible beauty as the plants age. One such tree is the Japanese umbrella pine, *Sciadopitys verticillata*, a beautiful evergreen conifer in the same botanical family as pines but significantly different in appearance. Japanese umbrella pine has needle-like leaves, as do pines, but the needles of Japanese umbrella pine are linear, flat, and many times wider and thicker than pine needles (Plate 205). Each needle of a Japanese umbrella pine is actually a double needle, two large, single needle structures fused together longitudinally at the center. The exotic foliage is lustrous and glossy and radiates upward around the stem like the ribs of an opened umbrella. A Japanese umbrella pine, its unique foliage glinting in the sun, is a striking sight in any landscape.

Sciadopitys verticillata is a native of Japan introduced into America in 1861. Like pines, Japanese umbrella pine can grow to be a large tree, as tall as 80–100 feet (24.8–21 m), but it is very slow growing and is rarely seen larger than 20–30 feet (6.2–9 m) in height and half as broad. The tree has a formal pyramidal shape that opens to a less formal habit as it attains maturity. Japanese umbrella pine grows so slowly as a young tree (as little as 6 inches [15.2 cm] per year) that it is rarely seen in mature form. Growth rate usually increases after 6 to 10 years to a more usual rate for conifers.

Japanese umbrella pine prefers sunny, protected sites, with afternoon shade in regions with high light intensity, such as the southeastern United States. It grows best in acid, fertile soils that are moist but well-drained. There are no serious disease or pest problems. The tree is hardy to zone 4. One of the finest specimens in the United States is to be found in the historic cemetery in Salisbury, North Carolina. This 35-foot (10.9 m) tree is now over 120 years old and inspires visitors with its classic, elegant beauty.

Sciadopitys verticillata can be propagated, with care and patience, by seed and cuttings. Seed is layered for 3 months in either moist sand at 63–70°F (15.5–19°C) or moist, acid peat at 32°–50°F (0–9°C). Seedlings take an extremely long time to gain significant growth and can take 4 to 6 years to grow 1 foot tall. As plants mature, however, their growth rate does increase. Cuttings taken in spring or midsummer can be rooted, but all cuttings are very slow to root. Rooting can be increased by soaking the cuttings in water to flush out some of the naturally produced resins in the tissue, and by applying rooting promoters. Treated cuttings placed on bottom heat under mist should root in about 6 months.

Because of the length of time it takes to reach a saleable size, either from seed or from cuttings, Japanese umbrella pine is not a common plant in the nursery trade. There is one cultivar, 'Pendula', an exceptionally graceful weeping form. Dwarf and yellow-foliaged seedlings are sometimes propagated and sold under various names.

Japanese umbrella pine is a very distinctive conifer—one not easily replaced with any other. The broad, straplike needles, whorled at the end of the stems, and the formal pyramidal habit make this a perfect specimen conifer, yet it is not hard to use, because of its handsome, muted green color. *Sciadopitys verticillata* is an excellent choice for corners of buildings and to define outdoor spaces such as decks and patios. However it is used, you'll be hopelessly captivated when you see the haunting foliage glistening up toward the sun. The only cure will be to bring one home, into your own garden, where its striking beauty can fascinate you for years to come.

Thuja plicata / Western red cedar   PLATE 206

At the North Carolina State University Arboretum, thousands of plants are brought in from around the United States and the world for evaluation as potential landscape plants. It is often surprising to learn which plants prove to be handsome and adaptable to the demands of the hot, wet southern Piedmont. Plants from similar climates that seem as though they should thrive may never be vigorous, while plants from dry and cool or other very different conditions may take off, surprising us all. One such plant is *Thuja plicata*, western red cedar, or giant arborvitae.

This richly handsome native of northwestern North America is adapted to entirely different, much cooler conditions than those of the southern Piedmont but it has developed into one of the most remarkable conifers currently in the Arboretum. The wood of this plant is exceptionally strong, making it an important lumber tree in its indigenous regions. The species is a large tree, reaching 50–80 feet (15.2–24.8 m) in height in the landscape but achieving extraordinary heights in its native habitat—on the order of 200 feet (62 m). It is uniformly conical and spreads to 20 feet (6.2 m) at the base, with rich green, graceful foliage that is tightly appressed to the branchlets. The foliage is reminiscent of other *Thuja* species but more elongated. Western red cedar is a stately tree that rises from the landscape like a lone mountain, covered with forests of emerald. The foliage of this plant has the added advantage of not browning out in winter like that of many of the forms of *T. occidentalis*, American arborvitae.

Western red cedar is cold-hardy to zone 5. It prefers, cool moist sites on well-drained soils that receive a great deal of moisture in full sun or partial shade but it has been an exceptional grower in the heavy clays at the North Carolina State University Arboretum, putting on 2–4 feet (0.6–1.2 m) a year in new growth. There are no significant disease or pest problems. Although bagworms have been seen on western red cedar, they do not appear to be nearly the problem that they are on Leyland cypress (× *Cupressocyparis leylandii*). Western red cedar can be propagated by seed or by rooting cuttings. Cuttings with some hardened wood are taken after significant cold winter weather and rooted under mist.

There are a number of exceptional cultivars of western red cedar. 'Atrovirens' is an excellent form for screens and hedges with especially bright, glossy foliage. 'Canadian Gold' is a broader form with dramatically gold foliage. 'Cuprea' is a dwarf form, reaching 3 feet (0.9 m) in height, with bright yellow new growth. 'Hogan' ('Fastigiata') is a tightly conical form with dark forest green foliage that remains beautiful all year. 'Pygmaea' is quite a small dwarf, to 2 feet (0.6 m), with bluish foliage. 'Semperaurea' is a rounded and compact form with foliage that turns gold in winter. 'Stoneham Gold' is an upright, medium dwarf form with bright yellow new growth. 'Stribling' is a dense, columnar form that grows straight upright. 'Sunshine' is a cultivar with an overall burnished yellow color on the foliage. 'Zebrina' (Plate 206) is perhaps the most common gold-variegated cultivar, with canary-gold bands across the sprays of foliage. It is one of the most delightful landscape conifers but the variegation tends to revert to green in the heat of the southern United States summer. 'Giganteoides' is a bionically fast growing cultivar, originally found in a seedling bed in Denmark, which is probably an accidental hybrid of *Thuja plicata* and *T. occidentalis.*

Thuja plicata has proven to be one of the most dependably attractive conifers in the Arboretum. It will make an excellent alternative screening conifer to Leyland cypress. The indigenous peoples of northwestern North America have used the trunks of western red cedar for canoes and totem poles. At the North Carolina State University Arboretum, western red cedar is an elegant and durable symbol of exceptional beauty and tough adaptability on the eastern border of the collections. It stands there as a living emerald totem—as tall against the southern sky as against the northwestern horizon.

Index